UNCERTAINTY IN NATIO[illegible]
POPULATION FORECASTI[illegible]

UNCERTAINTY IN NATIONAL POPULATION FORECASTING

Issues, backgrounds, analyses, recommendations

Nico Willem Keilman

SWETS & ZEITLINGER B.V. AMSTERDAM / LISSE

SWETS & ZEITLINGER INC. ROCKLAND, MA / BERWYN, PA / PUBLISHING SERVICE /

Library of Congress Cataloging-in-Publication Data

Keilman, Nico.
Uncertainty in national population forecasting : issues, backgrounds, analyses, recommendations / N.W. Keilman.
p. cm.
Includes bibliographical references and index.
ISBN 9026511418
1. Population forecasting--Statistical methods. I. Title.
HB849.53.K45 1990
304.6'01'12--dc20 90-45023
CIP

CIP-gegevens Koninklijke Bibliotheek, Den Haag

Keilman, N.W.

Uncertainty in national population forecasting : issues, backgrounds, analyses, recommendations / N.W. Keilman. -
Amsterdam [etc.] : Swets & Zeitlinger. - (Publications of the Netherlands Interdisciplinary Demographic Institute (NIDI) and the Population and Family Study Centre (CBGS), ISSN 0923-3512 ; vol. 20)
Ook verschenen als proefschrift Utrecht, 1990. - Met index, lit. opg.
ISBN 90-265-1141-8
SISO 302 UDC 314 NUGI 601
Trefw.: demografie

Cover design by Harmine Louwé - Studio Dunbar
Printed in the Netherlands by Koninklijke Wöhrmann B.V., Zutphen

ISBN 90 265 1141 8
NUGI 601

PREFACE

It is easier to predict how the next population forecast will look like, than to make a reliable forecast of the population itself. Experiences of the last few decades have shown that national populations are very difficult to foresee, and a high degree of uncertainty is present in long-term demographic forecasts. In this book I describe the results of an investigation into the uncertainty connected with population forecasts at the national level, as they are produced by statistical agencies.

Various persons have contributed to this study. Frans Willekens has been a continuous source of inspiration for issues related to demographic forecasting, demographic methods, and demographic theories. Without him, the research project would not have been undertaken. Henk Becker provided many useful suggestions regarding relevant problems in the fields of futures studies and philosophy of science.
My first experiences with population forecasting were when I worked at the Netherlands Central Bureau of Statistics, Department for Population Statistics. Rob van der Erf, Rolf Verhoef and Hans van den Brekel introduced me into this fascinating part of demography. Harri Cruijsen, Joop de Beer, Hans Sanders, Peter Meyer and Rik Zakee were not only excellent colleagues, but after I left the CBS and started my work at the NIDI, they showed continuous interest in the project. Also, I am grateful for the support I received from the CBS in the form of supply of data and access to archives.
At the NIDI, many persons have contributed directly or indirectly to the project, and I gratefully acknowledge their help: Peter van de Mortel, Rudi Schreurs, and Peter Ekamper. Discussions with Evert van Imhoff, Jan Bartlema and Jan Schuur have clarified my thinking. I have appreciated Nicky Williams' assistance with the English text. Jacqueline van der Helm and Joan Vrind did an excellent job in preparing several drafts of manuscripts and typescripts.
Discussions with various colleagues were extremely useful: Thomas Baumgartner, Helge Brunborg, Gustav Feichtinger, Janina Jozwiak, Tomás Kucera, Anton Kuijsten, John Long, Siv Schéele, Jan Schoonenboom, Tore Schweder, Per Sevaldson, Inger Texmon, Harrie van Vianen, Ad Vossen, and Guillaume Wunsch.

A final word of thanks goes to my wife Nina, who shaped the conditions under which a project with severe time constraints can be completed: she gave a perfect example of "voorwaardenscheppend beleid"!

July 1990

TABLE OF CONTENTS

1. INTRODUCTION

1.1 | Background of this study

The poor performance of demographic forecasts has been demonstrated repeatedly. Less well known, however, are the background and the consequences of this apparently limited predictability of population developments. In this book I attempt to explore several issues connected with the uncertainty which seems to be inherent to demographic forecasting in general, and, in particular, to official national population forecasts as regularly produced by statistical agencies. How large is this uncertainty? What are the reasons that it occurs? How is it handled in the actual practice of national population forecasting? How do the users of demographic forecasts cope with this uncertainty? Is the uncertainty different for different demographic components? Is it possible to discover any regularity and structure in the uncertainty connected to forecasts of population trends? How could a demographic forecaster reduce uncertainty? The underlying thought is that any efforts to reduce the uncertainty in population forecasting requires a detailed analysis of the extent and sources of this uncertainty (Stoto, 1988).

While taking up these and related questions, I accept the fact that demographic forecasts are inherently uncertain, and that this uncertainty is inextricably bound up with population forecasting. Demographic processes have a low degree of predictability[1]. This position is analogous to the influence which Polish-French mathematician Benoit Mandelbrot had in the 1960s on the way to view the problem of noise in telephone lines which are used for the transmission of information between computers. The traditional technique to drown out noise was to increase the signal's current. But there was the annoying fact that some spontaneous noise could never be eliminated. Mandelbrot, applying Cantor sets to the analysis of transmission errors, argued that instead of trying to increase signal strength to drown out more and more noise, engineers should settle for a modest signal, accept the inevitability of errors and use a strategy of redundancy to catch and correct them (Gleick, 1987, 92). This view is deeply rooted in the philosophy of science, and it dates back to the 1920s and 1930s. Werner Heisenberg formulated his famous Uncertainty Principle in 1926: the product of errors in position and momentum of an electron is constant. Therefore, when an electron's position is measured accurately, we can't tell what its speed is, and vice versa. Stated more generally, important aspects of reality cannot be known with certainty, let alone be predicted. Parallel with Heisenberg's ideas (or was it mere coincidence?), Kurt Gödel proved in 1931 that in any system there will always be statements which can neither be proved nor disproved. Thus, for certain statements it will be a matter of belief, not of knowledge, to decide whether they are true or not. Heisenberg's Uncertainty Principle and Gödel's Incompleteness Theorem tell us that "...there is a limit to what we can do, no matter how clever we are, or how hard or for how long we try..." (Churchhouse, 1985, 90). Similarly, I accept the fact that the demographic aspects of human behaviour (childbearing, living arrangements and life styles, health and morbidity, migration etc.) can only be explained to a small extent, let alone be predicted satisfactorily. Therefore, forecast errors are bound to occur.

The view that the limited predictability of certain processes is inherent to the phenomena involved, and not merely a consequence of our (hopefully temporary) ignorance, has slowly gained acceptance in such diverse fields as meteorology, population biology, economics, but also in specialized studies such as that into the dripping faucet, or the growth of a snowflake. Nevertheless, in demography the current paradigm is that the poor record of demographic forecasting is mainly due to the lack of knowledge about demographic aspects of human behaviour. The task to which I have assigned myself in this book is to demonstrate that, to a certain degree, uncertainty in demographic forecasting should be accepted, and can be accepted. This does not imply, of course, that I deny the relevance of studies which attempt to explain the demographic behaviour of individuals or groups. The point is that we should also (and much more so than in the past) study the uncertainty which accompanies population forecasting.

A number of factors account for the fact that, until recently, forecast uncertainty has so rarely been the subject of research in demography (Keyfitz, 1984a, 12).

Firstly, the time needed for forecasting and evaluation is relatively long in demography, because population trends evolve relatively slowly. Some 20 years or longer are necessary for a thorough study of the reliability of such forecasts. Meteorologists only need a week in all to make a forecast, to compare it to reality, and to update the forecast. For economists, a period of between a few months and a year or two is sufficient. But population forecasters may need as long as a generation: 20 to 30 years. *Simulated* forecasts (postdictions) could provide some insights, but these are dangerous for a proper assessment of uncertainty as I will argue in section 1.3.

Secondly, demographic forecasts, particularly those at the national level, are often officially authorized. Together with the fact that a relatively small number of demographers are concerned with future population developments, this means that competing population forecasts, each of which possibly shows a different demographic future, are lacking.

Thirdly, forecast users are not used to working with uncertainty. They want

point estimates: future population statistics (which form the basis of their planning) in the form of a single figure. They do not like working with interval estimates: a whole range of possible outcomes, for instance in the form of a probability distribution. The fact that population forecasters never offered the users the possibility of working with interval estimates cannot be seen apart from the user's attitude discussed here.
Finally, many users of the outcomes of population forecasts are specialized in a particular field, such as housing, regional policy, transport, energy, economics etc. For them, uncertainty regarding future population trends is often far less important than the uncertainty in their own specific field.

In recent years, however, a growing interest can be noted in issues connected with the accuracy, predictability and uncertainty of population forecasts. Three reasons can be given: the record of national demographic forecasts is sufficiently long, planners have developed planning concepts which can deal explicitly with uncertainty of the environment, and the forecasting discipline has matured.

Firstly, in the 1950s, many industrialized countries started to produce a set of regular national forecasts. The first evaluations of those forecasts, which were carried out in the 1960s and 1970s, led to the conclusion that the predictability of population trends was fairly limited, when traditional demographic techniques were used. Initially, the reaction of demographers was to investigate historical trends with more sophisticated demographic instruments (cohort analysis, parity-specific analysis) and to explore future population trends by means of other approaches (Delphi-approach, scenario writing, stochastic cohort-component models, ARIMA models). Nevertheless, the success of these new approaches, measured in terms of the accuracy of the resulting demographic forecasts, was still limited. Sharply declining birth rates were observed in most industrialized countries during the 1960s and 1970s, and these trends were insufficiently foreseen by demographers. This led, in the early 1980s to a systematic empirical assessment of the record of demographic forecasts. Important contributions are those of Keyfitz (1981) and Stoto (1983), who analysed the ex-post observed error in the population's growth rate in a large number of countries. Ahlburg (1982) investigated the accuracy of birth forecasts produced by the United States Bureau of the Census. These demographic studies were published shortly after Ascher (1978) devoted a chapter to the accuracy of forecasts of the population size of the United States in his influential book "Forecasting - an appraisal for policy makers and planners". These empirical studies which were carried out around 1980 were not the first ones. Early contributions, scattered over the literature, are those of Dorn (1950), Myers (1954), White (1954), Grauman (1967), Macura (1968) and Siegel (1972). But before 1980, none of these studies led to a recognition of aspects of uncertainty in official national demographic forecasts. The disclaimer following the article on the accuracy of demographic and household projections in the US by

United States Bureau of the Census (USBC) official Jacob S. Siegel (1972, 67) is most illustrative in this respect: "The opinions expressed here do not necessarily reflect the position of the U.S. Census Bureau." It reflects the common opinion at that time that national agencies did not have a firm viewpoint regarding forecast uncertainty. But shortly after 1980 the situation changed. Since 1984, the USBC analyses systematically ex-post observed errors of their forecasts (USBC, 1984; Long, 1987; Long and McMillen, 1987; USBC, 1989). The Netherlands Central Bureau of Statistics (NCBS) uses the average of ex-post observed errors in live births in historical forecasts to set the margin between the high and the low fertility variant in the forecasts of 1984 and 1988, and it investigates the pattern of errors in age structures (NCBS, 1985a, 19, 33; NCBS, 1989, 17, 31). Statistics Sweden devotes a section to errors in historical forecasts in its publications concerning the national forecasts with jump-off years 1986 and 1989 (SCB, 1986, 76-81; SCB, 1989, 60-63). Bretz (1986) investigates ex-post observed errors of national demographic forecasts in the Federal Republic of Germany. Without a long series of forecast errors, observed over a period of 20-30 years, such studies would not have been possible (see also Henry, 1985, 55).

The second reason for an increased interest in the issue of uncertainty in population forecasting is connected with concepts of planning. In the 1970s, a shift can be noted from comprehensive planning to strategic planning. Baxter and Williams (1978, 68), Willekens (1984, 357ff.) and Vossen (1988, 49ff.) discuss the link between population forecasting and planning.
Comprehensive planning was the major planning concept in the 1960s and early 1970s. It can be characterized as positivistic. The emphasis was on rationality and comprehensiveness, and on long term developments. But its poor record in the early 1970s led to a radically different planning concept. Strategic planning emphasizes flexibility. Its primary aim is no longer to achieve a priori formulated goals, but rather to start with a preliminary goal and, during the implementation of the plan, to reconsider this goal and decide whether it appears appropriate. The planning activities focus on developing a strategy, and regular information-gathering is important (Capon and Hulbert, 1985, 124). Short-term goals are precise, whereas long-term goals are to be obtained in the planning process. Decisions are broken down into short-term commitments, which are periodically updated as new information becomes available. Uncertainty in the environment of the plan's object, for instance relevant future demographic trends, can thus be taken into account. This is not to say, however, that the majority of the users of demographic forecasts appropriately handles the uncertainty which is connected with these forecasts. Some users simply ignore uncertainty and gamble on having initially chosen the correct forecast. Given the record of national population forecasts, this can hardly be considered as leading to good policy-making. Others work in a planning situation where the uncertainty of the demographic environment is of far less importance than that in

their own field. For instance, a company producing popular records requires for its investment decisions future numbers of the population aged 15-30 years, for the next five years, say. The uncertainty in these numbers is much less than that connected with the future spending power and musical taste of these young adults. Therefore demographic input to the investment plans can very well be taken in the form of point predictions, instead of probability intervals.

Finally, the science of forecasting has recently focused on the assessment of forecasting performance (Makridakis, 1986). During the late 1960s and 1970s the field of forecasting started to become popular. Makridakis gives three reasons: the appearance of commercial econometric services; the influence of statisticians such as Kalman, Box, and Jenkins; and the widespread use of the computer. He labels the 1970s and 1980s as a learning ground for forecasters. Initial forecasting successes were followed by large errors. As a consequence, the evaluation of forecasting performance was brought into the forefront of research activities (Makridakis, 1986, 17).

1.2 | Objectives and problem formulation

The objectives of this book are:

- to describe the sources of the uncertainty which is connected with national demographic forecasts, and the reasons why it emerges;
- to quantify that uncertainty by means of an evaluation of historical forecasts;
- to explore possible regular patterns in ex-post observed forecast errors, and to attempt to link these errors to social, organizational and demographic background factors;
- to give an estimate of the uncertainty to be expected in future population forecasts; and finally,
- to give recommendations as to how the accuracy of future population forecasts may be enhanced.

These objectives will be pursued by answering the following research questions.

1 How is uncertainty handled in the practice of present day population forecasting?
2 How large is the uncertainty which is connected with results (output) and assumptions (input) of national population forecasts? What is the accuracy of stock variables in a forecast, such as the total population size or the population broken down by age and sex? And what is the accuracy of flow variables: numbers of live births, deaths and migrants? How does the accuracy of these forecast results compare to that of indicators which are used to set up the assumptions: total fertility rate, life expectancy etc.?
3 Do ex-post observed errors of population forecasts exhibit any regular

patterns? What is the relation between these errors on the one hand, and, on the other hand, social and demographic developments, changes in forecast approach, and the organizational structure in which these forecasts are produced?

4 How large are the errors which may be expected to occur in future population forecasts?

5 Which recommendations may be given to statistical agencies in order to deal effectively with the uncertainty of population forecasts? How might these agencies reduce the uncertainty in their future forecasts?

An important consideration when formulating the questions under item 2 is the following. There is a certain logical hierarchical order in forecast results and forecast parameters. Between the highest level of total population size, and the lowest level of parameters (age-specific demographic rates), one finds the age- and sex structure of the population, and the components of change. When accuracy at a certain level is viewed as a result of the accuracy at lower levels, this opens up possibilities for an explanatory analytical-demographic analysis of forecast accuracy of the entire forecast process (Kuijsten, 1989, 7).

The questions formulated under item 3 need some clarification.

Generally speaking, research in the field of population forecasting serves, among other purposes, to increase the accuracy of these forecasts. Therefore it is important to know what the contribution is of forecasting methodology to this accuracy, as compared to the contribution generated by general social and by more specific demographic circumstances. Answers to questions as formulated under item 3 can be used for the programming of research in the field of population forecasting. For example, if forecast errors turn out to be influenced mainly by real trends in demographic indicators that are observed after the forecast was produced, and much less so by a choice for more advanced forecast methodologies, then monitoring the observed trends should have a much higher priority for forecasters than the construction of more sophisticated models and methods.

1.3 | Concepts and definitions

1.3.1. Forecast, projection, prediction

Although agreement exists in scholarly demographic journals about the distinction between a *forecast* and a *projection,* the current practice of statistical agencies isn't generally in conformity with the usual interpretation of these notions. Therefore, let me refer to the definition which is provided in the IUSSP Multilingual Dictionary, and which I shall adopt throughout: "Population projections are calculations which show the future development of a population when certain assumptions are made about the future course of population change, usually with respect to fertility, mortality and migration. A population forecast is a projection in which the assumptions are con-

sidered to yield a realistic picture of the probable future development of a population" (IUSSP, not dated, 90). From this definition it follows that projections are *conditional,* developing the consequences of the assumptions that are made. A forecast is *unconditional:* based on current scientific insights, a forecaster gives his best guess of what the future population will be. Tomorrow he may have a different opinion, but today's forecast reflects what is currently considered as a plausible future.
The general practice of statistical agencies deviates from these definitions. For instance, the Netherlands Central Bureau of Statistics (NCBS) wrote about its 1970 forecast: "A forecast is ... in no way a prediction, but merely the determination of the consequences of certain assumptions" (NCBS, 1971, 13; my translation). This is illustrative for what statistical agencies generally do: they produce forecasts, but they call their products projections. It is left to the user to judge the plausibility of the assumptions (see also Brass, 1986, 4, who supports this view). Most agencies do not clearly state whether their assumptions are realistic or not. The reluctance of official forecasters to admit that they aim at providing a realistic picture of the future population is understandable, given the relatively poor record of population forecasting[2]. Nevertheless, it is not justified in my opinion, for three reasons. Firstly, forecasters are in a much better position to judge the plausibility of a given set of assumptions than the average user (a trained demographer is not considered an average user here). If the forecaster is not able to tell whether his calculations paint a realistic picture of future demographic numbers, who else could be? Secondly, the average user simply acts as if the statistical agency had produced a forecast in the sense defined earlier. He cannot assume that the agency based its calculations on a set of assumptions which are not very likely. Thirdly, this reaction is unnecessary. In chapter 2 I argue that forecast variants and forecast target years are appropriate instruments to reflect the ex-ante uncertainty in a demographic forecast.
It will be clear that a sharp distinction between a forecast and a projection cannot be given in a certain situation. When one considers a broad range of *possible* futures, there will be a gradual change from *probable* (plausible) to *unprobable* futures. Moreover, different forecasters will attach different degrees of plausibility to the same extrapolation of a certain indicator.

It should be noted that in the 1980s some statistical agencies have shifted the interpretation of what they produce towards the accepted definition of a forecast: examples are the NCBS and the Czechoslovak statistical agency (Keilman and Kucera, forthcoming).
The discussion above shows the reasons why I prefer to speak of a forecast, unless it is explicitly assumed that the calculations do not show a realistic picture of the future population (for instance because currently observed demographic indicators are kept constant for the future, or because it is assumed that fertility will reach replacement level within 10 years).
The investigations will be limited to *official national demographic fore-*

casts, and most empirical evidence relates to post World War II forecasts of the population of the Netherlands, as produced by the Netherlands Central Bureau of Statistics. A number of studies deals with forecast accuracy for subnational areas, for instance White (1954), Henry and Gutierrez (1977), Isserman (1977), Schéele (1981), Stoto and Schrier (1981), Voss et al. (1981), Palit et al. (1982), Müller (1982, 1987), Openshaw and Van der Knaap (1983), Brunborg (1984), Hämäläinen (1987), Murdock et al. (1987), Smith (1987) and Rideng (1988). Kuijsten (1989) gives a systematic review of most of these studies. There are two reasons to limit this study to forecast accuracy at the national level. First, for reasons to be described in section 1.3.3, the empirical investigations focus on ex-post observed forecast errors. Only a long series of historical forecasts would open up the possibility of drawing firm conclusions regarding, for example, a possible improvement of accuracy over time. Such a series exists for national forecasts, but not for forecasts at the regional level. Second, at the lower regional level the variability due to internal migration is usually so large, compared to the variability due to other demographic components, that not much can be said about accuracy in fertility, mortality and international migration. At the same time I am aware of the fact that some national forecasts simply result from aggregation over the regional dimension of results of a multiregional forecast. In that case regional aspects of forecast uncertainty should be considered as well. The national demographic forecasts of the Netherlands (for which empirical analyses will be carried out) are produced by the Netherlands Central Bureau of Statistics, while the Ministry of Housing, Physical Planning and Environment is responsible for the regional and local population forecasts, the latter types of forecasts being hierarchically subordinate to the national forecast. Therefore the present investigations may be viewed as a first step towards a comprehensive analysis of forecast errors in national, subnational and categorical forecasts.
I have restricted the analyses to official demographic forecasts as they are routinely being produced by statistical agencies in industrialized countries. Two arguments have guided this choice: (i) official demographic forecasts, as opposed to products of individual researchers, are widely used in government planning; (ii) the institutional base of these forecasts guarantees a certain continuity in the production process of the forecast, which makes it worth while to formulate recommendations with respect to forecast uncertainty.

The word *prediction* is seldom used in demography, much less often than in economics and econometrics. I shall refer to a prediction as a statement about the future value of a variable, based on a calculation with an explicit model or a verified theory. The difference with a forecast is that a forecast is based on relatively poorly developed behavioural theories, at least in formalized form. Indeed, cohort-component models, which are routinely used by statistical agencies to produce a population forecast, do not contain behavioural relationships between the variables. They consist only of definition equations.

1.3.2. Uncertainty, accuracy, error, risk, and ignorance

Uncertainty will be interpreted as *degree of forecast inaccuracy,* and operationalized as *forecast error.* Schéele (1981, 3) states that a forecast error is observed uncertainty; uncertainty refers to the future, a forecast error (at least in the ex-post sense, see below) refers to the past. In this book I adopt simply a definition of *uncertainty* to refer to a lack of predictability about structure, outcomes or consequences in a decision or planning situation, cf. Hertz and Thomas (1983).

Rademaker (1979) distinguishes seven degrees of uncertainty in statements about the future.

1 The statement is impossible and inconsistent with present knowledge.
2 The statement applies to phenomena that are inherently unknowable: metaphysics, mysticism or religion is involved.
3 The statement applies to phenomena which are presently unknowable, but profound research may disguise (some of) the unknown in the future.
4 The statement is not a priori incorrect, and even plausible. An exhaustive argumentation cannot be given, yet it may guide action. It is a conjecture, a guess, or a strong belief.
5 The statement has been tested successfully several times. It is a hypothesis, an attempt to explain observed phenomena, and further testing is necessary.
6 The statement has generally been accepted, and it is part of the existing body of knowledge. It belongs to a comprehensive set of verified theorems.
7 The statement is a tautology, or absolutely valid otherwise.

Some forecasters argue that they operate at level 3 (they call their products "projections"), but in section 1.3.1 it was argued that this is an incorrect interpretation. Demographic forecasters do in fact give statements about future demographic developments at levels 4 and 5, and uncertainty in demographic forecasts may arise for several reasons. First, an estimate of the relevant demographic variable (for instance, the number of births in 1996) for some future year may be given, but its actual value may deviate from its expected value because of the randomness of the process involved. This is the *probabilistic* interpretation, in which uncertainty is conceived as a measure of concentration of some probability distribution (Mishan, 1979, 338; Uffink, 1990, 6). Knowing the probability distribution means that one can assign probabilities to outcomes. Some authors, particularly those dealing with cost-benefit analyses, speak of *risk* in this case (Dasgupta and Pierce, 1978, 176; Becker and Dewulf, 1990, 13). Second, there may be insufficient knowledge of the processes which guide human behaviour. In that case, uncertainty is equivalent to *ignorance.* "Uncertainty, strictly interpreted, implies ignorance of the probability distribution." (Mishan, 1979, 339). Becker and Dewulf (1990, 13) give the following illustration of the difference between risk and uncertainty: mortality forecasts by conventional

causes of death are risky, those due to AIDS are uncertain. In this second interpretation, the answer to uncertainty is more research into human behaviour. A third interpretation of uncertainty in demographic forecasts concurs with the second one. The argument is that *demographic processes are inherently unpredictable,* because human behaviour cannot be explained. Boudon (1986) is the most prominent one in his analysis of theories of social change. His conclusion is that individuals have a variety of possible actions, and this makes processes describing human behaviour essentially unpredictable. In other words, Boudon argues that definite "laws" are in principle beyond the reach of theories of social change, a view supported by Henry (1987, 10) in the case of demography. In chapter 3 I give an extensive discussion of this third argument.
It should be noted that the definition of uncertainty adopted here - lack of predictability - covers all three interpretations given above.

I want to stress that, although in the classification above, ignorance is explicitly attached only to uncertainty in the second sense, one may argue that ignorance causes uncertainty in all three interpretations: it is just the *degree* of ignorance which differs between the three types of uncertainty. With uncertainty in the first sense, we know the expected value of the variable for which a forecast is made, for instance the total number of births in 1996, but not the time at which an individual women gives birth to a child: this is a random variable. The probability distribution of this random variable is the same for all women involved. When the variance of the distribution is large, the actual number of births will most probably differ much from the expected value (given a certain number of women involved). We know this expected value, and the ignorance is expressed by the variance of the distribution. With the second type of uncertainty, we have no knowledge of the distribution. Thus we cannot give an expected value. However, with proper research we will be able to discover this distribution, which is the same for all women. With uncertainty in the third sense, there is no hope of discovering "the" distribution. Women have a variety of possible actions, which is expressed by the fact that different women have different distributions (or, perhaps, the same distribution, but with different expected values or other parameters). It is logically impossible to infer the probability distribution of the time of childbearing from one single event, and thus there is complete ignorance. In practice, all three types of uncertainty and of ignorance will be encountered in a given situation, of course.
For practical reasons, I shall speak of ignorance only in the second sense of uncertainty, but the choice is rather arbitrary.

Because ignorance plays a role in all three types of uncertainty, it will be clear why some authors attach ignorance to the first interpretation and view ignorance as the source of probability. As discussed for instance by Suppes (1984, 46, 83), Laplace was extremely clear in stating that for him probability arises from ignorance of true causes. The same attitude was implicit in other writings in the eighteenth century (Simpson, Lagrange, Gauss), and it persisted in the nineteenth century. Determinism, as opposed to probability, would be possible when all causes are known and their effects can be predicted. Indeed, the real test of determinism is predictability (Suppes, 1984, 31).

The focus on forecast accuracy and forecast errors necessitates a discussion on forecast appraisal and the usefulness of forecasts. Many authors have emphasized that the accuracy of a forecast is just one aspect of its usefulness, and in many instances not the most important aspect. Other aspects of forecast usefulness, which may be of greater value than accuracy, relate to the impact of the forecast and how well it can be used for appropriate decisions and policies. Examples of such other factors are the level of detail of the information set, the costs of application, internal consistency, clarity of assumptions and the extent to which it can legitimate action (Ascher, 1978, 3; Steece, 1982, 466; Long and McMillen, 1987, 172; Long, 1987, 17; Bartlema, 1987, 23). One should *use* medium and long-term forecasts, but not necessarily *believe* them (Hogarth and Makridakis, 1981, 129; Bartlema, 1987, 23). Capon and Hulbert (1985, 131) refer to a statement which was formulated within the context of a certain company's strategic planning: "... a key point is to be more concerned with perspective than prediction (in the sense of an accurate forecast, NWK)...". Makridakis (1984, 1) goes one step further by stating that most criticism on forecasting can be traced back to too high expectations from the users. Yet accuracy is a focal point in the usefulness of forecasts because it contributes to the credibility of forecast producers *and* policy makers, because it is much easier to measure than other aspects of usefulness, and because it is the only major criterion for which general propositions about the impact of various factors on forecast performance can be made. It should be noted, however, that for future studies in general, the notion of accuracy is problematic in two cases:

- the case of self-fulfilling and self-defeating projections, which become inaccurate just because they are effective (Ascher, 1978, 5; Keyfitz, 1984b, 17). But these types of projections generally are of minor importance in national population forecasting, because demographic behaviour of individuals is relatively independent of the results of population forecasts at the national level;
- the case of conditional projections, for which one cannot assess the accuracy of the projections per se, without an evaluation of the conditions at the same time (Ascher, 1988). This problem relates to the definition of a projection, and the distinction between a projection and a forecast. Precise definitions of these concepts were given in section 1.3.1. There I argued why most explorations of the demographic future which are produced by statistical agencies are, or can be interpreted as, unconditional forecasts. It is stressed here once more, that forecasts are unconditional. Because an unconditional forecast simply asserts what the future population will be, it can be checked after the demographic event by noting how close it came

to the actual figure. This kind of checking has little meaning for conditional projections, because the (non-demographic) conditions should be evaluated (see also Keyfitz, 1984b, 17).

Forecast uncertainty is used here in the sense of forecast accuracy, i.e. the probability that the forecast comes true. Accuracy, thus defined, is complementary to *precision* (Van Dantzig, 1952, 197). The precision of a forecast is reflected in some measure of the width of the "forecast interval" (cf. confidence interval in statistics), that is the set of all states or events which would make the forecast become true. By widening the interval of a forecast, and thus decreasing its precision, one may obtain a highly accurate forecast. But an imprecise forecast will be less informative than a precise one. This poses a dilemma to the forecaster who uses a traditional deterministic approach, in which no probabilities are assigned to the forecast interval. However, if one works with *confidence intervals for forecast results* the dilemma disappears. Such an interval quantifies the complementarity of accuracy and precision. A two-thirds (67 per cent) confidence interval is narrower than a 95 per cent interval.

Forecast uncertainty could, in principle, partly be studied by *rerunning old forecasts with altered parameter values.* (Such an approach would give insight in forecast errors due to wrong parameter values, not in those due to model misspecification. These two types of forecast errors are discussed in more detail in section 2.1.) But I think this approach is to be avoided, because uncertainty will be underestimated. "... Forecasting ... that something will happen after it has happened ought to be much easier than forecasting the same event before it happens, when it is still in the future. Once the event is past all the attendant circumstances and statistics of the economic and social conditions and covariates are known. The data resources for those variables that act on population are infinitely more extensive and more accurate than we have at our disposal before the event, when both the demographic statistics and the series that will affect them are unknown or can only be estimated with very large errors ..." (Keyfitz, 1985a, 60).
If one would not restrict the investigations to the old approach (supplemented with new parameter values) but, more generally, study the forecast errors resulting from a different approach, the situation would be somewhat different from that described in the previous paragraph. The cohort-component method may be replaced by an entirely different method, for instance a time series extrapolation. If this other method would be entirely objective, in the sense that it would result in the same extrapolations regardless of who carries it out, then one may expect to be able to answer questions such as "How accurate would those old forecasts have been if they would have been based on a different methodology?". However, there is so much subjectivity in the extrapolation methods which I know (even automatic Box-Jenkins procedures require an assessment of patterns and critical levels of key program parameters) that I am not confident that this would be good practice.

The analyses in this book are directed towards *long-term accuracy* of demographic forecasts. Errors which show up in a single forecast are not considered, as they would relate to *short-term accuracy.* This choice is guided by a major objective of the present study which was described in section 1.2, namely to discover possible regularities (in the sense of time series) and structures (in the sense of explanatory factors) in forecast accuracy and forecast errors. This requires a series of forecasts. To evaluate the accuracy of one single forecast is an activity connected with *monitoring*. In chapter 3 I discuss this and several other activities when I review the stages which can be distinguished in the production of a forecast.

1.3.3. Ex-post and ex-ante errors

Forecast uncertainty is appropriately reflected by *ex-post observed forecast errors,* and much less so by *ex-ante errors*. Errors of the latter type are generated by a stochastic model, which admittedly provides much more insight in error structures than a mere set of observed forecast errors. The problem with ex-ante errors, however, is that in the stochastic model, one or more parameters *have* to be constant in time, whereas in reality this may not be the case. Ex-post errors capture the latter variability as well. Also, some important factors with great impact on the demographic variables may have been left out of the forecast. In an ex-post analysis one covers these factors as well, albeit in implicit form only. One can expect, therefore, that a model-based approach which results in ex-ante errors indicates a higher degree of accuracy than an empirical approach in which the focus is on ex-post errors. Pflaumer (1988, 141) finds that a 90% confidence interval for the future population size of the United States, resulting from a stochastic population model (and thus reflecting ex-ante errors), is narrower than the interval defined by the high and the low variant of the 1984-based forecast of the USBC. Next he concludes that "...the U.S. Bureau's presentation of high and low projection series probably leads to an overestimation of the uncertainty in population forecasting...". However, this conclusion is not warranted, because an analysis of ex-post observed forecast errors of growth rates (empirical approach) demonstrates that the high and low variant of the USBC correspond approximately to a two-thirds confidence interval for the total population (Stoto, 1983, 18). Kuijsten (1989) applies Pflaumer's approach to the Netherlands, and he finds two-thirds confidence intervals for total population size (p.43) which are approximately 50% narrower than the range defined by the high and low variant of the forecast of the Netherlands Central Bureau of Statistics (NCBS, 1987, 15). (A small part of the difference is due to the fact that the NCBS assumed an annual net-immigration of 10 to 34 thousand individuals for the years between 1986 and 2000, whereas Kuijsten ignored international migration.) Similar to the case of the United States, empirical investigations of ex-post observed errors have shown that the high and the low variant of the official forecast of the NCBS represent the margins of an approximate two-thirds probability interval of total population size (NCBS, 1985a, 32; Keilman,

1986, 70). Hence, a comparison of ex-ante errors and ex-post errors of total population size shows very clearly that ex-ante errors tend to underestimate forecast uncertainty. This justifies the choice for ex-post errors as a means to study forecast uncertainty. Therefore, when I use the term "(forecast) error", this should be interpreted as ex-post observed error, unless explicitly stated otherwise.

Nevertheless, it should be noted that the choice for ex-post errors is not without problems. Long and McMillen (1987, 150) mention three problems when forecast errors are used to measure accuracy. It requires a long series of forecasts, for which the results can be compared to actual demographic trends; the method's sophistication is likely to have changed over time, and more so the longer the series of forecasts is; and inaccuracy is due as much to variability of population trends as to characteristics of the forecast method. These problems are dealt with in the present study as follows. I have collected relatively long series of forecast results since 1950 for several demographic indicators. Moreover, in the analysis of chapters 4-6 I focus on changes in forecast methodology and the likely consequences for forecast accuracy, and I apply a statistical model which separates the impact of ex-post observed population trends on forecast errors from that which relates to characteristics of the forecast method.

1.4 | Plan of this book

In chapter 2 I review a number of issues related to uncertainty in demographic forecasting, mainly based on existing literature. The discussion in section 1.3.2 on the sources of uncertainty in demographic forecasting is used to make a typology of errors (section 2.1). A large number of measures to evaluate forecasts have been mentioned in the literature. In section 2.2 I discuss the most widely used measures, together with some criteria which may guide the choice for a proper measure. Suggestions as to how to express uncertainty in official national population forecasts are given in section 2.3. In spite of the fact that statistical agencies apply a strictly deterministic approach for their population forecasts, a number of instruments exist which can be used, even in such an approach, to express forecast uncertainty, and some of them are actually used. An important instrument is sensitivity analysis, and in section 2.4 I review new results of a sensitivity analysis which was recently carried out for the multidimensional demographic projection model in general, and that one including marital status and a two-sex algorithm in particular.

The aim of chapter 3 is to formulate hypotheses on uncertainty in demographic forecasting, which are to be tested in subsequent chapters. In section 3.1 I analyse the social and political context in which forecasts are produced, and section 3.2 contains a description of the stages of the cyclical process which results in a population forecast. The discussions in sections 3.1 and 3.2 are geared towards an identification of those aspects of the

wider context of population forecasting which are likely to contribute most to forecast errors. An important issue is to what extent behavioural theories contribute to the accuracy of demographic forecasts, and in section 3.3 I discuss the most important aspects involved. The analyses in sections 3.1-3.3 result in section 3.4 into the formulation of a number of hypotheses on accuracy and errors in national demographic forecasts.
Chapters 1-3 discuss several aspects of uncertainty in national population forecasting in industrialized countries in general. In chapters 4-7 I focus on experiences for and empirical evidence taken from the Netherlands. Chapter 4 presents a description of the methodology which was used in the national population forecasts in the Netherlands since 1950. I consider number and type of components, period and cohort orientation, the introduction of variant projections, as well as the contribution of social demography to national forecasting, and the organizational context in which forecasts in the Netherlands were produced. This analysis results in the identification of important turning points in the methodological history of post-World War II population forecasting in the Netherlands.
In chapter 5 I give a first descriptive analysis of ex-post observed forecast errors in the Netherlands since 1950. In an analysis of forecast errors, a thorough discussion of the quality of the collected data is essential. Therefore, several issues connected with the data (consistency checks, estimation of missing information, interpretation and definitions) are reviewed in section 5.1. Next I present in sections 5.2-5.4 forecast errors in total population size, annual population growth, the age structure, and components of change (fertility, mortality, international migration). The chapter concludes with a summary of findings (section 5.5), where I try to answer the following questions. Are recent forecasts more accurate than older ones? How fast do errors grow as we look further into the future? And how fast do they diminish as we approach a certain, initially chosen, year?
The answers given in section 5.5 can only be tentative, because the period in which ex-post errors are observed, and the length of the forecast period have confounding effects on forecast errors in a series of forecasts. Therefore I describe and apply in chapter 6 a model which disentangles these confounding effects. A formal separation of the two effects facilitates a much more profound analysis into the problem whether recent forecasts are more accurate than older ones than could be given by means of the descriptive analysis in chapter 5. Section 6.1 gives a discussion of the problem of confounding effects, and in section 6.2 I describe an additive (linear) separation model which builds on insights obtained earlier from age-period-cohort models (APC models). In this model it is assumed that observed forecast errors are composed of three effects: a duration effect, which accounts for the increase of uncertainty when we look further into the future; a period effect, which relates to the observation period; and a forecast effect which summarizes the average error connected with the forecast which was produced in a certain year. A solution to the identification problem inherent to models of this kind is proposed, and estimation aspects are

discussed. Section 6.3 contains results of the application of the model to errors in fertility, mortality and international migration. For the first two components, errors in absolute numbers as well as crude rates are analysed. First I estimate the model for fertility and mortality for the entire period 1951-1986, and next for the shorter period 1976-1986. Thus I am able to study the stability of the general conclusions which emerge from this chapter for the choice of the period of observation. International migration is analysed in terms of errors in net numbers of immigrants for the smaller subset of observations only. Finally, I demonstrate in section 6.4 how the accounting equation of demographic growth can be used to relate model results for errors in numbers of births, deaths and immigrants to results for errors in annual population growth.
The purpose of chapter 7 is to explore possible links between ex-post observed forecast errors and potential explanatory factors. This will be done in the form of an explanatory model. Explanatory factors were identified in section 3.4, partly on the basis of theoretical considerations, but also based on empirical findings. Sixteen hypotheses were formulated in section 3.4, and with the available data I can test nine of them.
In section 7.1 I discuss the models to be used and the dependent and independent variables. To enhance interpretation, separate models are formulated for the absolute value of the error and its sign. The former type of models are linear, and they are estimated with Ordinary Least Squares. The latter are Probit models and these are estimated with Maximum Likelihood techniques. Section 7.2 presents results for absolute errors in crude birth rates, in the life expectancy of males and females, and in immigration and emigration. In section 7.3 I select several models which are used to generate out-of-sample predictions of forecast errors for the period 1984-2000.
Confidence intervals around forecast results give an appropriate refelection of the ex-ante uncertainty which the forecaster expects to be connected to the forecast. A confidence interval consists of three elements: an upper boundary, a lower boundary, and a probability that the actual figure will lie between these two boundaries. Thus, the concept of a confidence interval is a logical extension of the notion of forecast variants, in the sense that it attaches a certain probability to the interval between the high variant and the low variant.In chapter 8 I review and apply two methods to construct confidence intervals around forecast results: (i) the method of Stoto (1983), which is used for total population size; and (ii) the method of Cohen (1986), which is appropriate for age structures and total population size.
Section 8.1 contains a brief introduction into the two methods for constructing confidence intervals. An application of these methods to total population size forecasts for the Netherlands is given in section 8.2. More detailed results on intervals around forecasts of a few selected age classes are presented in section 8.3.
Conclusions are formulated in chapter 9. The main findings of this book are summarized in section 9.1, in which I refer to the objectives and the research questions which were formulated in section 1.2 of the present

chapter. A distinction is made between general findings, which immediately apply to national population forecasting in general, and empirical findings which were derived from the data that I collected for the Netherlands. The latter empirical findings, which include the test results of the hypotheses, may only be generalized to other countries when the situation in these countries is deemed comparable with that in the Netherlands. Section 9.2 contains some recommendations for the reduction of uncertainty in demographic forecasting at the national level, and suggestions how national statistical agencies may adequately handle the uncertainty which is inherent to their forecasts. Finally, in section 9.3 I discuss some questions which are open to further research.

NOTES

1) Some authors distinguish between uncertainty and risk. I shall treat these two notions as equivalent, see section 1.3.
2) To change the name of the product is one possible reaction to the large errors observed in demographic forecasts. Another reaction is to produce something entrirely different. An example of the latter strategy is the Bariloche study, which is a normative projection, reflecting the political aim for larger international justice. This study was an answer to the investigations by Forrester and Meadows for the Club of Rome, which were rather technical-economic unconditional forecasts. Demographers are bound to produce forecasts, because policy-makers need insight in the likely future. This is not to say, of course, that normative demographic projections are superfluous.
3) In certain cases, international migration may be the reason why a projection becomes self-defeating. If, in addition, this type of migration has a strong impact on the country's demographic development, it is useless to investigate the accuracy of this projection.
4) The reason is that ex-ante errors are often based on models for which the underlying assumptions apparently do not hold. A notable exception is given by Cohen (1986), who fits a stochastic model for total population size to Swedish data for the period 1780-1980. He finds that confidence intervals around his point estimates are of approximately the same width as those generated by the method of Stoto (1983), for a forecast period not longer than ten years. The methods of Stoto and Cohen are used in chapter 8, where I draw conclusions about the uncertainty in current national demograhic forecasts in the Netherlands.

2. SOME ISSUES IN UNCERTAINTY IN DEMOGRAPHIC FORECASTING

In this chapter I review some issues related to uncertainty in demographic forecasting, mainly based on existing literature. Several sources of uncertainty in demographic forecasting are discussed, which results in a typology of errors (section 2.1). A large number of measures to evaluate forecasts have been mentioned in the literature. In section 2.2 I discuss the most widely used measures, together with some criteria which may guide the choice for a proper measure. A number of suggestions as to how to express uncertainty in official national population forecasts are given in section 2.3. Statistical agencies apply a strictly deterministic approach for their population forecasts. Nevertheless, a number of instruments exist which can be used, even in such an approach, to express forecast uncertainty, and some of them are actually used. I discuss in section 2.3 variants, target years, and the mode of presentation of the forecast. An important additional instrument is sensitivity analysis, and section 2.4 contains a review of new results obtained in sensitivity analyses of multidimensional demographic projection models in general, and those including marital status and a two-sex algorithm in particular.

2.1 | Types of forecast errors

Which types of errors arise in population forecasts? Hoem (1973) and Keyfitz (1977, 227) were among the first authors to discuss this issue. Berk and Cooley (1987, 257) review types of errors in forecasting social phenomena in general. Based on these analyses, I distinguish seven types of errors. The frame of reference is the cohort-component method, which is the standard approach employed by statistical agencies for forecasts at the national level (see also section 2.3.1).

1 Errors in observed trends Due to the difference between the "de jure" and the "de facto" population, population registers, and thus the demographic data derived from these registers, do not accurately reflect actual behaviour. Moreover, if population registration is absent, or not complete, basic data used in forecasts may have to be derived from sample surveys. This is likely to create sampling errors, non-response errors, etc.

2 *Errors in the jump-off population* The production of population statistics based on population registration is never 100-percent accurate. So, with time, the population numbers determined by the statistical bureau will increasingly deviate from the figures given in the population registers. Without a system of population registration, the jump-off population will have to be estimated from census data and additional vital statistics.

3 *Rounding errors* Rounding errors are made in observed demographic rates, but also in the results of population forecasts.

4 *Randomness in parameters* Occurrence-exposure rates (o-e rates) form the larger part of the model parameters for a population forecast. An o-e rate is defined as the ratio of the number of events (such as childbearing, death, emigration, marriage) to the number of person-years the population is exposed to the risk of experiencing the particular event. Observed o-e rates for fertility, mortality and the like are composed of a trend plus a random fluctuation. Population forecasts, however, are based on an extrapolation of the (supposed) trend, whereby randomness is not taken into account.

5 *Errors in values of exogenous variables* Ex-post analyses may show that the extrapolation of exogenous variables (in cohort-component models these are birth rates, death rates, etc.) has resulted in an over- or underestimation of observed trends.

6 *Sudden shifts in parameters* External circumstances such as war, disasters, serious economic depression, and the like, but also sudden changes in legislation or technology can dramatically influence the course of demographic parameters.

7 *Inaccurate model specification* The population model used might not be entirely valid. When simulating population trends over a given period with the aid of such a model and observed data, the model results should, ideally, be exactly the same as the actual trends. However, this is not always the case.

Errors of types 1-3 may be called measurement errors. Errors due to randomness in parameters are consistent with what I called in section 1.3.2 the probabilistic interpretation of uncertainty. The two other interpretations of uncertainty, viz. uncertainty due to ignorance and inherent uncertainty are responsible for errors of types 5-7, but it is often impossible to make a clear distinction between these two sources when errors of either type 5-7 are observed.

Errors in observed trends (category 1) and errors in the jump-off population (category 2) are related to the quality of population registration. In most developed countries this does not give rise to serious problems, although it should be noted that the quality of the jump-off population gradually deteriorates as the time at which the last census was taken (or any other independent check) lies in the more distant past. In countries where registration is substandard or even absent, one must resort to data from surveys, in addition to censuses. Indirect estimation techniques can then give rise to inaccurate estimates for the basic demographic trends. Rounding errors (category 3) are unimportant, unless the forecast involves very small populations at the local scale. Various studies have shown that randomness in parameters (category 4) does not lead to serious forecast errors (Pollard, 1966; Sykes, 1969; Feichtinger, 1971; Schweder, 1971; Le Bras, 1974; Cruijsen and Van Hoorn, 1983a; Brunborg, 1984).

The most common forecast errors arise because of a misspecification of the values of exogenous variables (category 5). Sudden shifts in parameters (category 6) can create large forecast errors, but they are not very common. For example, if it has been proven that an economic recession encourages women to postpone childbirth, and if such an economic slump does indeed occur, and persists, then the resultant errors in birth rates are usually classified under the fifth category of errors. In the past, model specification errors (category 7) were caused in particular by the fact that international migration was not included. For small countries this resulted in forecast errors which were quite important for migration-sensitive ages (up to about 40 years), as compared to the errors made for other age groups.

2.2 | Measures to evaluate forecast accuracy

A key question to be answered when a demographic forecast is evaluated, is how to *measure* the accuracy, or, in accordance with the definition of accuracy employed here, how to *measure ex-post forecast errors*. The literature gives a large number of measures: mean error, mean percentage error, mean absolute error, root mean square error, Theil's U, to name a few. Ahlburg (1982), Armstrong (1985) and Pflaumer (1988b) discuss several measures. Dammers (1990) reviews 30 accuracy measures, and presents a checklist for 21 of them, to help in the selection of an appropriate measure of accuracy. (Nine of the original 30 measures can be replaced by others.) This checklist is an extension of a much shorter list presented by Armstrong (1985, 355). Timmermans and Borgers (1984) give a descriptive evaluation of 30 measures (partly overlapping with those of Dammers), and Op 't Veld et al. (1984) check the analytical performance of these 30 measures using simulated data.

Dammers divides his 30 measures in five groups (which are not all mutually distinct): "simple" measures (for instance, the deviation between observed and forecast value, or the mean error, or the mean absolute error), relative measures, which correct for the units in which the relevant variable is expressed (for example, mean percentage error), quadratic measures (such as the root mean squared error), proportional measures (for instance, the ratio of observed to forecast value) and measures for a wrong sign (overestimation or underestimation). The aspects which are considered for each measure are the following (see also Somermeyer, 1961; Baydar, 1983; Kuijsten, 1984):

- characteristics: is the measure *homogeneous;* does it increase *monotonously* with increasing difference between observation and forecast; is it *bounded;* does it have a *simple interpretation*?
- applicability: is it appropriate to evaluate the estimate of *one number,* or of *a series of numbers;* can we apply it to a forecast expressed as *a level, a change, a growth rate, a proportion, a negative number;* can the measure be applied to a *ratio variable, an interval variable, or only to an ordinal variable*?

Timmermans and Borgers (1984) distinguish between two types of measures: descriptive measures and statistical measures. Descriptive measures simply indicate in one number the degree of similarity between two sets of values. In this sense, all measures discussed by Dammers are descriptive. They do not allow for significance tests. In contrast, statistical measures are devised to allow conclusions regarding the hypothesis that the observed and the forecast data stem from the same population. Most of the statistical measures are either related to the χ^2 (χ^2, ϕ^2, likelihood ratio statistic), or they stem from information theory (information index, information gain, etc.).
The conclusion of Op 't Veld et al. (1984, 21ff.) is that statistical measures possess a number of disadvantages, the most relevant one for this study being the fact that they cannot be used when structural zeros are to be expected. Quadratic measures have the disadvantage that they are very sensitive to outliers.

In the analyses of chapter 5, three measures are widely used: the mean percentage error (MPE), the root mean squared percentage error (RMSPE), and a modification of Theil's U_2, with the following definitions:

$$\text{MPE} = 100\ N^{-1} \Sigma\ (F_t - O_t)/O_t,$$

$$\text{RMSPE} = 100\ [N^{-1} \Sigma\ \{(F_t - O_t)/O_t\}^2]^{1/2}, \text{ and}$$

$$U = \{\Sigma\ (F_t - O_t)^2 / \Sigma\ (A_t - O_t)^2\}^{1/2}.$$

In these three expressions, F_t and O_t represent the forecast value and the observed value at a certain point in time, respectively. N is the total number of forecast or observed values which is evaluated. A_t is an alternative "forecast" or benchmark "forecast", of which the quality is compared to the actual forecast, because much of the accuracy may be due to persistence and inertia in demographic trends (Keyfitz, 1977, 230). As an alternative "forecast" one could take the results of a naïve extrapolation in which the relevant variable is kept constant at the level which was recently observed. When U exceeds one, this means that the alternative "forecast" was more accurate than the actual forecast. U is below one in the opposite case, and U=0 indicates a perfect forecast. U as defined here differs from Theil's U_2 in two respects (Theil, 1966; Armstrong, 1985). Firstly, the alternative "forecast" At is not contained in the definition of U_2. Secondly, O and F (and A) refer to *levels* here, whereas they represent *changes* in Theil's U_2. Therefore, U_2 compares the root mean squared error (RMSE) of the forecast to that of a no-change "forecast", when O is taken to denote actual change, and F forecast change. U_2 may very well be used for the evaluation of a forecast of stock variables, for instance total population size. But *changes* in stock variables as such are seldom analysed by demographers - rather they investigate *components of change*: births, deaths, and migrations. In the following chapters I study the accuracy of these components, and therefore I need an explicit assumption on the value of an alternative "forecast". This explains why I use A_t in the definition of U.

In chapter 5 the alternative "forecast" for a certain variable is the value as it is observed in the year prior to the jump-off year of the relevant forecast. This rule is applied to the evaluation of forecasts for the number of live births, the crude birth rate, the crude death rate, the life expectancy, the number of immigrants and the number of emigrants. For the number of deaths this would be a forecast which is "too naïve" for a country like the Netherlands with its rather strong population growth between 1950 and the mid-1970s. In that case the naïve "forecast" consists of a crude death rate which is constant in time. Numbers of deaths are obtained by multiplication with the forecast value of total population size. (This approach is unsatisfactory in the case of a population which not only grows, but also ages. The reason is that the crude death rate corrects for population size, not for age composition. The standardized death rate and the life expectancy *do* control for the age structure, but these indicators would require a much more elaborate analysis. The accuracy of the life expectancy will be investigated for a few selected forecasts in section 5.4.2.)
A positive MPE reflects, on average, an overestimation of actual numbers, and vice versa. Positive and negative terms compensate each other in the MPE. When overestimations have as much impact for the forecast user as underestimations, the MPE is not appropriate, and the RMSPE is a good alternative. Smith and Sincich (1988) argue that the MPE reflects the "bias" of the forecast, and that its accuracy in a strict sense can be evaluated using measures like the RMSPE or the mean absolute percentage error.
When the three measures MPE, RMSPE and U are checked for their characteristics using Dammers' checklist, we see that they are all homogeneous of order zero (i.e., when both observed and forecast values are multiplied by the same factor, the measures remain unchanged), they are *not* a monotone

function of the difference $F_t - O_t$, RMSPE and U have a lower bound of zero (MPE has no lower bound, none of the three measures is bounded above), they are all easy to interpret, and they can be applied to summarize in one figure the errors in a series consisting of such ratio variables as levels, changes, and proportions. In chapters 6 and 7 I use the simple difference F_t-O_t, as well as its absolute value $| F_t - O_t |$ to analyse and model the error in forecast variables in distinct years. These two measures are monotone, but not homogeneous of order zero. $| F_t - O_t |$ has a lower bound of zero. In chapter 7, finally, I use the *sign* of the error $F_t - O_t$ to study the possible structure in overestimations and underestimations in demographic forecasts.

2.3 | How to express forecast uncertainty in cohort-component methodology?

Despite of the fact that statistical agencies apply a strictly deterministic approach for their population forecasts, a number of instruments exist which can be used, even in such an approach, to express forecast uncertainty, and some of them are actually used. In this section I discuss the concepts of forecast variants, of target years and target cohorts, and I review the manner of presentation of uncertain forecasts. An important instrument to express uncertainty is sensitivity analysis, and in section 2.4 I review new results of a sensitivity analysis which was recently carried out for the multidimensional demographic projection model in general, and the one including marital status and a two-sex algorithm in particular.

2.3.1. Variants

A survey on national population forecasting among statistical agencies in 30 industrialized countries, conducted in early 1988, showed that the cohort-component approach was used by all agencies (Keilman and Cruijsen, forthcoming). Although other approaches exist, e.g. economic-demographic models, time-series models, microsimulation models and judgemental approaches (Long, 1984; Land, 1986), the cohort-component method is the standard for forecasting at the national level. This method is entirely deterministic. Therefore it is important to investigate possibilities of expressing uncertainty in the current practice of national population forecasting: how do producers of population forecasts deal with uncertainty, and what is the situation for the users of these forecasts?

As early as 1933, Thompson and Whelpton published several variants of their estimates of the future population growth in the United States because they were not sure about the course the demographic components would take. Ever since, this has become common practice among forecasters. But there is no general agreement as to the interpretation of forecast variants (Long, 1988). One interpretation is that there is only one demographic future for the population under consideration, for which one set of *qualitative* assumptions can be formulated. However, uncertainty is introduced when these qualitative assumptions are translated into *quantified* values of demographic input rates for the projection model, and this uncertainty is expressed by more than one variant. Such variants are called "uncertainty variants" or "probability variants" in this interpretation. A second interpretation of the notion of forecast variants is that of "alternative futures". In this case the gap between two extreme variants is much larger than in the first interpretation. These two different interpretations of the notion of forecast variants are neatly illustrated by the fact that among 18 industrialized countries with more than one fertility variant in their national population forecast, ten exhibit a relatively narrow range between the high variant and the low variant, whereas five countries show a wide range (Keilman, 1990). The relative range was defined as the difference between the values of the Total Fertility Rate in the high and the low variant, as a percentage of the simple average of the two extreme TFR-values. The first group, consisting of ten countries, displays a relative range in the year 2000 of between 8% and 20%, with an average value of 13%. The five countries in the second group have relative ranges of between 32% and 45%, and the average value is 40%. I assume that the narrow gap countries interpret their variants as uncertainty or probability variants, whereas the other countries probably prefer to speak of alternative futures, when they consider their variants. It should be stressed that the uncertainty variants interpretation is consistent with the definition of a forecast, which expresses a realistic future population trend. When the agency publishes more than one demographic future for its country, the question remains which set of calculations should be picked as the forecast. From now on, variants will be interpreted as uncertainty variants, unless stated otherwise.

Forecast variants may be formulated for each component: fertility, mortality, international migration, and (in case marital status is included in the projection model) marriage and marriage dissolution. The choice of the components for which more than one assumption should be formulated is guided by the following considerations (NCBS, 1982, 17).

1 The significance of each individual component for the results of the forecasts. For instance, the future number of children aged 0-4 is much more sensitive to changes in fertility than to changes in mortality or migration. But migration affects the future number of males aged 20-24 relatively strong.
2 The past course of the components. The more erratic the observed developments, the more difficult it is to estimate the future course of the component in question.

The first consideration may be used when a sensitivity analysis has been performed, see below. The second consideration results from our ignorance about the quantitative relations between the demographic components and the underlying social, economic, cultural, political and technological fac-

tors. Here, we are interested in quantitative answers to questions such as:
- how do processes of female emancipation influence the number of children a woman will bear?
- how is mortality affected by technological developments?
- to which extent will the demographic behaviour of ethnic minorities presently living in many Western countries differ from that of the total population?

The *qualitative* answers to these and similar questions (e.g. mortality will slowly decline due to improved medical technology, and in spite of unfavourable environmental developments) are the same for each variant. Only one demographic future is sketched in the forecast assumptions. It is their *quantification* which introduces uncertainty, to be expressed by more than one variant.

The distance between corresponding variables in the high and the low variant is indicative of the degree of uncertainty of the component in question. This implies two things:
- as we look further into the future, uncertainty grows, and the range between high and low variant should grow accordingly;
- there are large differences among components in range between variants.

The fact that uncertainty increases with time can be expressed by making use of target years or target cohorts, see 2.3.2. Differences among components in range between variants are illustrated by the results of the 1988 Survey on National Population Forecasting in Industrialized Countries. Among seven countries with more than one mortality variant, the relative range in the (male) life expectancy at birth, as assumed for the year 2000, varies between 1% and 6%, and it is 3% on average (Keilman, 1990). Thus the gap is much narrower for mortality than for fertility (13% on average, see above). This reflects the common belief among demographers that it is easier to give a realistic extrapolation for mortality than for fertility. (External migration shows relative ranges from 29% to 600% in the year 2000. However, the latter range was defined with the aid of the net-number of immigrants, and therefore the interpretation of this range is problematic.)
In a new forecast, the distance between the high and the low variant for a certain parameter in a particular year can be set on the basis of ex-post forecast errors of the relevant parameter in historical forecasts (De Beer, 1988).

2.3.2. Target years and target cohorts

The more distant the future, the greater the uncertainty. In population forecasts this can be attenuated in two different ways. Firstly, by an increasing gap between a high and a low variant. Secondly (and this can also be done for components which are expressed by only one variant), by defining a so-called target year. A target year is a year in the future, until which an explicit development is assumed in the demographic indicator to be extrapolated. In the target year, the indicator attains its ultimate value, and it remains constant thereafter. The reason is that in the forecaster's perception it is barely

possible to select, for a so remote period, a plausible trend from a large range of possible future trends. In practice, target values are determined using past trends, survey data, data from more advanced countries, substantive reasoning etc. Intermediate values are obtained by interpolation between recent trends and the target value. For instance, the NCBS uses a third degree polynomial (also known as two points Hermite interpolation) with the conditions that the interpolated curve fits to the value most recently observed, its slope during the past few years, the target value and to a zero slope in the target year (NCBS, 1984, 81).
The distance between the jump-off year of the forecast and the target year expresses the degree of uncertainty. Different components may be given different target years. Instead of target years, one may also work with target cohorts for certain components.
Figure 2.1, taken from the NCBS forecast of 1984, illustrates the notion of a target year or cohort. In that forecast, the year 2000 was chosen as the target year for fertility and first marriage. For mortality and divorce the NCBS-forecasters did not want to assume an explicit trend beyond 1995, whereas international migration remained constant already after 1990. The cohort approach which was applied during the extrapolation of fertility and first marriage resulted in a much more regular trend in these components than that observed in divorce and remarriage, which were analysed by a period approach.
It should be noted that between 1990 and 2000, the calculations could no

Figure 2.1. Total fertility rates and completed cohort fertility rates, the Netherlands, observed and according to 1984-based forecast of the Netherlands Central Bureau of Statistics (NCBS)

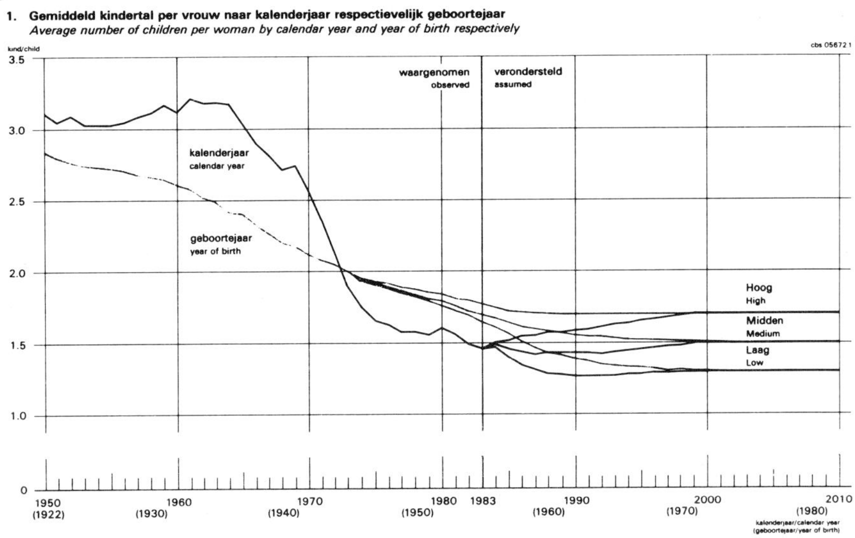

Source: Cruijsen (1985, 33)

longer be called a forecast, but they gradually turned into a mere projection. As early as 1928, Whelpton used target values and target years for his extrapolation of death rates, but the target years coincided with the final year of his forecast (Whelpton, 1928, 257). Because the extrapolation period covered 50 years (1925-1975), Whelpton cannot have used the target years and target values to express uncertainty. Long and McMillen (1987, 147) present target cohorts for fertility in the USBC population forecast.

2.3.3. Presentation of uncertain forecasts

When the performance of a forecast is poor, this may induce two types of reactions from the forecaster (see also note 2 of chapter 1). The first one is to produce something different: instead of forecasts, one may wish to calculate mere projections. Demographers cannot do this, because forecasts are necessary for planning purposes. The second type of reaction is to give the product a different name, and this is what many demographers actually did. They started to call their forecasts "projections", or "illustrative projections". In my view, this is not a good practice. Observed levels of uncertainty have to be accepted for the time being, its magnitude has to be estimated, its sources have to be analysed, and the presentation of the forecast should reflect forecast uncertainty (Page, 1982, 455; Alho and Spencer, 1988, 5). Forecast users have to make their decisions in the face of at least some uncertainty. In an ideal situation the user is provided with a confidence interval: a range of possible outcomes together with relative likelihoods. This would make the user aware of the fact that he must not apply a planning methodology which assumes that the forecast is very precise. It is of great importance that decisions which are to be based on a population forecast are robust against forecast errors, or else they should be sufficiently flexible so that they can be adjusted to revised forecasts (Baxter and Williams, 1978, 68).

Bandel (1990) presents results for local forecasts in the Stockholm area in the form of so-called *quality tables* (see also Bandel and Schéele, 1990). A quality table gives the expected error (overestimation or underestimation) in a certain forecast result. For a particular forecast which Bandel presents (e.g. the future population in family houses), the expected error is a function of age and of forecast duration. The assumption is that the average error observed in historical forecasts equals the expected error in the present forecast. As an example, consider table 2.1, which is taken from Bandel (1990, Appendix 1).

The observed error is Schéele's S, defined as $S = | F-O | / \sqrt{O}$, see Schéele (1981, 5) and Dammers (1990, 365). F and O represent the forecast value and the observed value, respectively, of a certain population subgroup. Schéele discovered empirically that an error measure thus defined is more or less independent of the size O of the observed population, and that it only varies with age and length of the forecast period. To illustrate the use

Table 2.1. Quality table for forecast results of children aged 0 living in family houses

	Length of forecast period (years)				
	1	2	3	4	5
	Mean observed error				
	1.094	1.441	1.621	1.743	2.253
Actual value	Expected absolute error				
10	3	5	5	6	7
20	5	6	7	8	10
50	8	10	11	12	16
100	11	14	16	17	23
200	15	20	23	25	32
500	24	32	36	39	50
1000	35	46	51	55	71
2000	49	64	72	78	101
5000	77	102	115	123	159
10000	109	144	162	174	225

Source: Taken from Bandel (1990, Appendix 1).

of the quality table, suppose for a certain municipality the future number of children aged 0 living in a family house three years from now will be 100. Then the forecast error can be expected to be 16 children. The value of 16 follows for the absolute error | F-O |, when S=1.621 and O=100 are inserted in the definition of S given above[1].

It should be noted that the probability of the absolute error | F-O | being, at most, equal to its expected value, is approximately 0.6. This result is based on the assumption that individuals follow the same binomial distribution for a given event, and that the sum (over all individuals) of a binomial distribution can be approximated by a normal distribution, see Schéele (1981, 7). Thus the quality table gives an approximate 60 per cent confidence interval.

The quality tables of Bandel and Schéele form an interesting innovation in the mode of presentation of uncertain forecasts. The minimum amount of information necessary to determine the interval around a particular forecast result F is the corresponding (that is, for the particular population category and forecast duration) expected value of S (see also note 1). The important question is, of course, if the mean observed error is equal to the expected error. Time series analysis of observed values of S can shed some light on this issue.

The more traditional way of communicating uncertainty to the user is to work with forecast variants. If several variants are included in the forecast,

they should be given equal weight in the presentation. Incorporating secondary variants in an appendix serves little purpose. The results of uncertainty variants ought, whenever possible, to be mentioned simultaneously. A statement such as "In the year 2000 the population of the Netherlands will most probably number 15.3 to 16.1 million persons" is much more effective in communicating forecast uncertainty to the user than "According to the medium variant the population of the Netherlands will number some 15.7 million people in the year 2000; the low variant puts the number at 15.3 million and the high variant at 16.1 million".
Uncertainty can be emphasized in tables by placing the variants side by side. By doing so the user is much more strongly confronted with forecast uncertainty than would be the case if three separate tables had been constructed. Hence one dimension of any table of forecast results should be reserved for forecast variants. In elaborate tables this is not always possible.
Placing forecast results in a historical context can also give the user an impression of the uncertainty involved. The greater the uncertainty, the more often one has to adjust the forecast. Cruijsen (1985, 40) illustrates this point nicely for the 1984 forecast of the NCBS: "...Only three years ago the population of the Netherlands in the year 2030 was estimated at 14.3 to 16.7 million. This forecast has meanwhile been put at 12.7 to 15.2 million. This example adequately shows that the value of long-term forecasts is limited..." (my translation). I can add that in the 1989 forecast the NCBS expects 14.3 to 18.0 million inhabitants (NCBS, 1990, 23).
As a further illustration, consider table 2.2, which contains the Netherlands' total population and its age structure in the year 2000 according to various forecasts made by the NCBS in the past.

Table 2.2 shows very clearly how uncertain forecasts for young age groups are. In a period of less than 20 years, the 2000-forecast for girls aged 0-4 had to be adjusted downwards from 1.179 million in 1965 to 480 thousand in 1984! This reduction of 60 per cent is equivalent to saying that in each year between 1965 and 1984, the expected number of 0-4 year old girls in the year 200 fell by 4.6 per cent. The uncertainty decreases rapidly with increasing age, and ages between 40 and 75 are fairly stable. However, at higher ages the variation increases again: for instance, between 1972 and 1984, rises of 17 per cent, 31 per cent and 64 per cent can be noted in numbers of women aged 75-79, 80-84 and 85 or over, respectively.
In chapter 5 a more detailed analysis of these and related phenomena will be given.

As far as the user is concerned, decision-making under uncertainty requires knowledge of "loss functions" or "cost functions", depicting the costs of a decision based on the wrong demographic inputs (Keyfitz, 1977, 233-235). Because population forecasting at the national level is general purpose forecasting (the latter concept will be elaborated in section 3.1.3), there is a multitude of users, each having their own cost function. Therefore a general

Table 2.2. Total female population and female population age structure in the year 2000 according to subsequent forecasts, the Netherlands

	Jump-off year						
	1965	1967	1970	1972*)	1975*)	1980*)	1984*)
				thousands			
All ages	10,543	9,037	8,579	8,082	7,674	8,060	7,887
0-4	1,179	794	697	564	484	528	480
5-9	1,037	754	673	581	518	535	479
10-14	940	726	668	585	522	512	455
15-19	883	688	660	580	477	480	433
20-24	834	646	649	600	430	466	445
25-29	759	601	600	522	511	562	545
30-34	653	570	571	574	582	623	613
35-39	586	586	580	583	591	627	624
40-44	554	554	548	551	559	589	587
45-49	530	530	524	528	536	556	553
50-54	556	556	542	543	548	562	560
55-59	424	424	406	407	411	423	422
60-64	365	365	347	348	353	362	365
65-69	345	345	321	323	329	336	342
70-74	315	315	287	288	295	303	315
75-79	277	277	244	244	254	267	286
80-84	172	172	147	147	154	170	193
85+	136	136	116	116	122	159	190

*) High variant.

review of cost functions will not be given here. It will suffice to state that cost functions are widely discussed in statistical and forecasting literature, but their use in the context of national population forecasting is probably extremely limited. Neither from the literature, nor from my own experience do I know any forecast user who actually deploys this concept. This may be due to the fact that demographic forecasts are presented without accompanying probabilities.

2.4 | Sensitivity analysis in multidimensional projection models

An important instrument to obtain insight in the uncertainty of deterministic demographic forecasts is sensitivity analysis. Large errors in the results of a demographic forecast may result from an incorrect specification of input parameters (age-specific occurrence-exposure rates for fertility, mortality, migration). Errors in input parameters (labeled as "type 3 errors" in section 2.1) are by far the most important source of inaccuracy in demographic forecasts. But not all occurrence-exposure rates are equally important in this respect. Those for fertility have greater impact on the population in the youngest age groups than those for mortality - mortality rates mainly affect higher age groups. Intermediate age groups are sensitive to international migration. The situation becomes more complicated when national population forecasts are produced with a model, in which individuals are not only broken down by age and sex, but also by an additional dimension such as region of residence or marital status. In that case we have to specify input values for internal migration rates and nuptiality rates as well. The more general model is known as the *multidimensional population projection model.* In this section I start with a general discussion of sensitivity analysis (section 2.4.1), and next I present some new findings for the sensitivity in multidimensional projection models (section 2.4.2).

2.4.1. Generalities

Sensitivity analysis is an instrument which can help to distinguish between important ("sensitive") and unimportant parameters (see section 3.2 for a discussion of the role of sensitivity analysis in the forecasting process). This can be done analytically or empirically. In the *analytical* method, differential calculus is applied to the projection model. If this were to become too complicated, simulation is used. In that case we speak of *empirical* sensitivity analysis. Input parameters are varied ("distorted") and results of subsequent projections are compared. In computing forecast variants, one also varies values of input parameters, but because the purpose of computing variant forecasts is not primarily to distinguish between crucial input parameters and less important ones, I do not interpret such variants as examples of sensitivity analysis (see also Pflaumer, 1986, 1988b). An interesting discussion on the link between results from variant forecasts on the one hand and ex-post observed forecast errors on the other is provided by Spencer in his report about the 1988 forecasts of the United States Bureau of the Census (USBC, 1989, 15). The root mean squared error in the rate of population growth was calculated for all USBC forecasts of the past 30 years. A comparison with the difference in growth rate between the high and the medium variant of the 1988 forecast, and one between the low and the medium variant of that forecast shows that the forecast roughly encompasses the historical error range for the first 15 years, and then it appears to widen. "...After 50 years the difference between the highest and lowest projected series represents a range of plus or minus twice the historical average root-mean-square error, or roughly a 95-percent confidence interval ... it should be pointed out that these figures are merely illustrative ... considerable work must still be done before a more reliable method of establishing confidence intervals is available ...".

Sensitivity analysis in the sense just described refers to the consequences of changes in *values of exogenous parameters* (birth rates, death rates, etc.). A different type of sensitivity analysis is to investigate the effects of a change of *model structure*. Brass (1986, 4) holds a plea for more systematic sensitivity analysis as an aid in chosing between various forecasting strategies. Such a sensitivity analysis is probably always empirical. It will not be discussed further here, as I limited the scope of this study to national population forecasts, for which the cohort-component approach is the dominant methodology.

The analytical approach to obtain sensitivity functions for multidimensional projection models was first applied by Willekens (1977). Later on, the multidimensional model was improved (see Willekens and Drewe, 1984; Keilman, 1985a), which necessitated new sensitivity functions. Ekamper (1990) investigated the sensitivity of the so-called linear specification of the multidimensional projection model, in which events are assumed to be uniformly distributed over the unit projection interval. An alternative to this linear model, namely the exponential specification of the multidimensional model has a number of advantages (see, for instance, Gill and Keilman, 1989; Van Imhoff, forthcoming), but it is not (yet?) used by statistical agencies.

Projection models are discrete-time models. Thus one could distinguish between the effect of a change in parameter value after one unit projection interval, and the effect in subsequent intervals. Cruijsen and Van Hoorn (1983b, 23) investigate both types of effects for the marital status projection model of the NCBS. Willekens (1977) analyses theoretical aspects of both types of effects. Arthur (1984) studies the consequences of perturbations in parameters of continuous-time models.

Still another aspect is the number of parameters for which the effect of a (simultaneous) change is analysed. Several parameters may be distorted at the same time. Arthur (1984) investigates this problem analytically in the context of continuous-time models. Cruijsen and Van Hoorn (1983, 27) follow an empirical approach.

Finally, one should consider which projection variable (output variable) has to be studied for the effects of a perturbation in input parameter. Stock variables (population by age, sex and possibly other dimensions, such as marital status) are the forecast results which are most commonly used for planning purposes. Hence they receive most attention in sensitivity analyses.

2.4.2. Sensitivity in marital status projection models with a two-sex algorithm

In this section I discuss some results obtained by Ekamper (1990) in his investigation of sensitivity functions for one specific type of multidimensional population projection model, viz. the *marital status projection model*. An important issue in such a model is the question whether it satisfies the so-called two-sex constraints: are male and female marriages equal during each projection interval? Are male divorces the same as female divorces? Does mortality of married men correspond with widowhood of married women? Ekamper (1990) derived sensitivity functions for a multidimensional marital status projection model with (except those defined by age and sex) four states (never married, currently married, divorced, widowed). His model includes an algorithm to solve the two-sex problem. This two-sex algorithm is currently used by the Netherlands Central Burau of Statistics; a technical discussion is given by Keilman (1985b). Van Imhoff (1988) shows that the algorithm belongs to a very general class of algorithms, based on weighted least squares optimization, which can also handle constraints between events in household models.

Ekamper (1990) applies matrix differentiation techniques to obtain partial derivatives and elasticities which measure the relative effect of a small change in a model parameter on the model variables. He investigates the following model parameters: age- and sex-specific rates for marriage, divorce, transition to widowhood, death and emigration, age-specific fertility rates and age- and sex-specific numbers of immigrants. Model variables analysed are population numbers broken down by age, sex and marital status, as well as total population. An empirical application with five-year age groups and a unit projection interval of five years is given for the Netherlands, in which observed demographic data for the period 1980-1984 and the actual population as of 1 January 1985 serve as input for a projection over the period 1985-1989. Selected results are shown in tables 2.3 and 2.4.

Table 2.3. Effects on output variables (stock variables) resulting from distortions in age-, sex-, and marital status specific mortality rates; selected mortality rates and output variables

Age group, sex and marital status corresponding with mortality rate			Age group, sex and marital status corresponding with output variable			Elasticity (per cent)
80-84	M	nev.married	85+	M	nev.married	-66
85+	F	nev.married	85+	F	nev.married	-43
70-74	M	married	75-79	M	married	-30
70-74	F	married	75-79	F	married	-15
80-84	M	divorced	85+	M	divorced	-87
80-84	F	divorced	85+	F	divorced	-53
85+	M	widower	85+	M	widower	-52
85+	F	widow	85+	F	widow	-45

The largest elasticities are found for mortality of unmarried persons in the oldest age groups. Table 2.3 shows for each combination of sex and marital status the highest elasticities when various age groups are considered. As can be seen from the table, a one per cent increase in the mortality rate for divorced males aged 80-84 reduces the number of divorced males at the end

Table 2.4. The effect of a one per cent increase during the period 1980-1984 in the marriage rate of never married males aged 20-24 on 1 January, 1980, on the projected population (by age, sex and marital status) as of 1 January, 1985

Age	Males				Females			
	never married	married	divorced	widower	never married	married	divorced	widow
				%				
10-14	0.0000				0.0000			
15-19	0.0002	-0.1892	0.0569	-0.0006	-0.0023	0.1645	-0.0112	0.0004
20-24	0.0258	-0.2194	0.0261	0.0714	-0.0814	0.2008	-0.0369	-0.0331
25-29	-0.4404	0.4180	0.4251	0.4368	-0.2772	0.1202	-0.0485	-0.0437
30-34	0.1704	-0.0587	0.0545	0.0847	-0.1502	0.0306	-0.0574	-0.0307
35-39	0.0628	-0.0167	0.0688	0.0840	-0.0581	0.0103	-0.0508	-0.0152
40-44	0.0351	-0.0113	0.0823	0.0780	-0.0336	0.0072	-0.0475	-0.0120
45-49	0.0126	-0.0068	0.0646	0.0552	-0.0127	0.0043	-0.0361	-0.0067
50-54	0.0066	-0.0045	0.0496	0.0340	-0.0059	0.0024	-0.0231	-0.0036
55-59	0.0045	-0.0033	0.0396	0.0252	-0.0027	0.0015	-0.0125	-0.0027

of the interval by 0.87 per cent. For married males and females, the effects of a distortion of the mortality rates are much smaller: 0.30 and 0.15 per cent, respectively (in the opposite direction).

Fertility is important for young age groups. Nevertheless, the effect of a one per cent increase of the fertility rate of married women aged 20-24 is that the number of children aged 0-4 at the end of the five-year interval rises with only 0.18 per cent. This elasticity of 0.18 per cent is lower than those for the effects which distortions in mortality rates generally have (on numbers of aged persons). For other ages of the mother the consequences of a change in the fertility rate on the number of young children are much smaller.

The two-sex algorithm adopted by the NCBS in its projection model possesses the nice feature that an increase in the marriage rate of males at a certain age has two effects. It *reduces* the marriage rates for unmarried males at other ages, and it *increases* the marriage rates for unmarried females at all ages (and vice versa for the sexes interchanged).

This so-called *competition effect* (Keilman, 1985b) is clearly illustrated in table 2.4, which shows the effects (expressed as elasticities) of a one per cent increase in the marriage rate of never married males aged 20-24 on the population by age, sex and marital status at the end of the projection interval. While such an increase raises the number of married males who belong to the same birth cohort by 0.42 per cent, it leads to a decrease of 0.22 per cent for married males aged 20-24 (at the end of the interval), and of 0.06 per cent for those aged 30-34. The higher marriage propensity for males aged 20-24 leads to higher numbers of married females at the end of the interval: plus 0.16 per cent for the age group 15-19, plus 0.20 per cent for ages 20-24, and plus 0.12 per cent for ages 25-29. Note the indirect effect of an increased marriage rate for never married males. Because more males marry, it leads, at the end of the five-year interval, to 0.43 per cent more divorced men and 0.44 per cent more widowers in the same birth cohort.

Occurrence-exposure rates other than those discussed in tables 2.3 and 2.4, for example those for remarriage, divorce, emigration and immigration have much lower effects on projection results. Their elasticities are generally below 0.1 per cent. Distortions in widowhood rates have effects which are comparable to those for mortality of married persons. Effects on the total population size of distortions in o-e rates are less than 0.01 per cent, and thus negligible. Simultaneous changes in more than one parameter, for instance a one per cent change in life expectancy at birth, would of course have more dramatic effects on the total population.

NOTE

1) In reality, one does not know the actual value O for a certain year in the future, but one only has a forecast F. Therefore, the absolute error $|F - O|$ should not be calculated on the basis of O (and S), but rather with a given F. In that case, the expression for the absolute error is

$$|F-O| = |\, 1/2\, S^2 \pm S \sqrt{F + 1/4\, S^2}\,|.$$

3. NATIONAL POPULATION FORECASTING AND SOCIETY: TOWARDS A THEORY OF FORECAST ERRORS

In this chapter I formulate hypotheses on accuracy in national population forecasting, to be tested in chapters 5-7. An analysis is given of the social and political context in which demographic forecasting takes place. Together with a detailed description of the subsequent stages of the forecast production process this leads to an ad-hoc theory of forecast errors.
I discuss in section 3.1 a number of issues relevant to the social and political context of forecasting and model construction in general. These issues are applied to national demographic forecasts in the Netherlands. An important aspect is the organizational framework in which the forecast is or was produced, and the notion of the "forecaster's institutional base". An analysis of possible sources of forecast errors must cover a much wider set of issues than just the technical problem of selecting an appropriate forecast procedure. This is not to say that those technical issues should be neglected. Therefore, section 3.2 contains a discussion of the six stages of the production cycle of the forecast that can be distinguished: 1. systems identification, 2. systems description, 3. model construction, 4. parameter extrapolation, 5. execution, documentation and implementation, and 6. monitoring. For each stage I discuss possible sources of forecast errors.
Keeping accuracy in mind, the most critical stages are those of systems identification, systems description and parameter extrapolation. The explanation of demographic processes plays a key role in the first two of these three stages, and the prediction of those processes is of paramount importance for the latter stage. I discuss in section 3.3 the reasons why demographic theories have failed to predict gross shifts in demographic trends in the Western world after the Second World War. This discussion, together with a review of the relevant empirical literature on forecast accuracy, leads in section 3.4 to a number of hypotheses on accuracy of, and errors in, demographic forecasts.

The theory of forecast errors to which I seek to contribute will consist of a set of conjectures, similar to theories of demographic behaviour. Moreover, it will contain generalizations describing "how things work", rather than "covering laws" in the sense of Hempel. Section 3.3 contains a discussion of several types of theories. The regularities I may discover consist of a set of more or less general trends, and some conditional "laws" of the form 'if A, then sometimes, or usually, B'. It is certainly not my intention to formulate general laws similar to the ones which exist in the physical sciences.

3.1 | The social and political context of demographic forecasting

In his opening address to the Fourth International Symposium on Forecasting (London, 8 July 1984) Sir James Ball suggested that up to fifty per cent of forecasting is art, and not science. Whether this holds true for forecasting in every field of application remains to be seen. But in the case of demographic forecasting one cannot deny that a great deal of inspiration and hence of subjectivity is involved, besides technical issues (projection model, extrapolation technique etc.). Long and McMillen (1987, 142) suggest that a projection represents the "science" side of the forecasting process, the "art" side to the process being the choice of "realistic" assumptions. But I think that science can be used not only for the projection, but also to guide the choice of the realistic assumptions.
Since there is art in this business, there is also an opening for societal influences, even more so than in the case where forecasting would be a matter of science only. The societal influences can range from the individual (a scientist's values, interest and ideology) to the social (professional orientation, group dynamics, the need of forecast users to justify their gambles) and to the political (would policy makers accept the assumptions behind and the results of the forecasts?). Therefore the social and political context of demographic forecasting has to be analysed if one wants to understand why some population forecasts go wrong, and others do not. Baumgartner (1987) has discussed a number of issues that are relevant for the analysis of influences from policy-making and socio-political context on population forecasting. His ideas originate in the record of energy forecasting of the past 20 years (Baumgartner and Midttun, 1987), and I shall apply them here to population forecasting. Thus I sympathize with Fildes' (1982) view, that an organization must be concerned with a much wider set of issues than just technical ones when it examines forecast performance. Although Fildes' analysis applies to a forecasting framework which is valid for forecasting in the private sector, many of the questions he asks find their parallel in the issues discussed in this chapter for the case of population forecasts produced in the public sector.

An important assumption is that the traditional rational model of decision making (sometimes called the "social-engineering perspective") is too sim-

plistic to take into account important aspects of the relationship between society and forecasting. In the rational model, goals are specified, alternative courses of action are identified for achieving these goals, these alternatives are evaluated, and the best alternative is selected. In the case of population, decision-makers in such fields as population policy, health policy, labour market policy etc. are provided with information regarding future demographic trends from a model representation of the population system.
The rational model presupposes pure efficiency, and it excludes traditional, emotional and political considerations. The information drawn from the model is based on a set of adequate data. The policy maker is unaffected by the population system. In this rational model the complexity and the conflicts between interest groups are underestimated, the authority and effectiveness of administrative and political organs is overestimated, and questions as to the sources of "inspiration" and "knowledge" are left out (Finsterbusch and Motz, 1980, 24; Baumgartner, 1987, 7; Baumgartner and Midttun, 1987, 12-13). A more realistic theory of forecasting should take a number of points into account.
Firstly, there does not exist just one 'correct' model (i.e. a model which strikes a proper balance between parsimoniousness, ex-post accuracy and contents) - models are always products of social interaction. Although the cohort-component approach is the standard in population forecasting at the national level, other approaches exist, e.g. economic-demographic models, time-series models, microsimulation models and judgemental approaches (Long, 1984; Land, 1986).
Secondly, there is not just one policy-maker who uses demographic forecasts. Users in the public sector and in the private sector form a highly diversified and heterogeneous group and they may even pursue contradictory policies. For instance, certain private firms in industrialized countries may prefer increased labour immigration in the nineties to cope with expected shortages at the labour market, whereas public authorities may prefer a decrease of immigration because of existing tensions on the housing market.
Thirdly, knowledge of social processes is limited and, to a large extent, it often defies quantification and registration. For example, processes of individualization and emancipation are believed to exert their influence on the life style of the individual, and hence on childbearing. In section 3.3 I shall come back to the limits of knowledge of social processes.

A more comprehensive perspective than the social engineering perspective should at least address three core issues regarding population forecasting and its socio-political context. What are the political aspects of data selection and technical/methodological questions? What is the relationship between modelling and the institutional and professional environment? And how are forecasting and political decision-making connected? These three issues will be taken up below.

3.1.1. Data selection and methodology

Within the framework of a purely descriptive demographic model, the choice for a cohort-component approach is logical. Cohorts are introduced because age is the most important demographic variable. And to split up population growth into its constituent components (fertility, mortality and migration) clearly enhances the understanding of population dynamics. Because of this logic, the broader question should be: why do demographic forecasters restrict themselves to purely descriptive, accounting-like demographic models, and why don't they work with socio-demographic or econometric models, which attempt to link demographic events or phenomena to explanatory factors? I assume that the dominant position of the descriptive demographic approach in national population forecasting cannot be seen apart from the existence of (some sort of) vital registration systems in Western countries. Data on the current structure of the population (by age, sex, region, and sometimes even more characteristics), and data on births, deaths, migrations, etc. are most often supplied to or produced by national statistical agencies. The primary aim of these registration systems is to provide information on the current demographic situation for administrative purposes. The system gives an answer to demographic questions of the type "How many?", but it does not tell why. Hence the cohort-component approach, which is essentially a one-to-one mapping of the demographic accounts contained in or derived from vital registration systems, does not allow the forecaster to include individual behavioural aspects in a direct way in his cohort-component model.

Now imagine a forecaster who wants to go beyond the detail defined by age, sex, region, marital status etc. Since vital registration systems are most often rooted in official legislation, it would be next to impossible to include more factual information (for instance, socio-economic status) into the registration system. But more important is the fact that "soft variables" (values, intentions, beliefs) would require a survey, since they cannot be included in a vital registration system which was created for administrative purposes. Many of these surveys have been organized in the past, and most of them aimed at a better understanding of childbearing, and its background. But in a discussion on the political and social context a number of questions arise (it should be noted that questions of this type apply more generally than to *demographic* forecasts and data collection). Is it the forecaster himself who decides whether such a survey will be organized? Does he have control over the key aspects (research questions, depth of analysis, sample size, variables, frequency)? A sample survey is an expensive and complex undertaking, which requires considerable financial funds, organizational skills, substantive insights and endurance. It cannot be assumed that our forecaster has complete authority to decide about the major aspects of such a survey. When working in a statistical agency, he has to compete with labour surveys, housing surveys, budget surveys etc. Perhaps he is able to raise external funds, but that introduces lines of influence from the funding organization. In short, there are many reasons to believe that the data collection pro-

cess is influenced in a number of ways because of political and social interests that go beyond the mere demographic ones.
An interesting example is provided by the case of the Netherlands. The first nation-wide fertility survey of that country was organized in 1969 (see, for instance, Moors, 1974). This was no less than 14 years after the first national fertility survey in the US - the Growth of American Families Studies of 1955. One important cause for this scientific lag was the fact that raising funds to finance such an undertaking was difficult in the Netherlands in the 1950s and early 1960s: Van Nimwegen (1981, *30) clearly showed how sensitive questions were about the background and motivation of childbearing in Dutch society in those years. Thus the data collection process may be influenced heavily by political and social factors, and the close link between vital registration and demographic forecasting has led to a dominance of the descriptive demographic approach, and thus of the cohort-component model, over other forecasting methodologies.

3.1.2. Professional and institutional contexts of population forecasting
The forecasters' institutional affiliations, their training and their professional characteristics are generally considered as important general aspects of the social and political context of forecasting (Keyfitz, 1984a, 20; Baumgartner and Midttun, 1987, 18). More particularly, they may even become potential explanations of why some forecasts are more accurate than others (Ascher, 1978, 7).

The aim of professional population forecasters is to do better than non-professionals, both in terms of methodology and accuracy. Most of them work in statistical agencies. Hence although they may have received training as a demographer, their orientation generally is towards numbers, tables and statistics and much less towards theories of social and individual behaviour. In their training, they have adopted a certain attitude towards data: a combination of respect and scepticism, and especially a concern for its integrity, as Keyfitz (1984a, 21) believes. In their forecasting organization, this orientation towards data is shaped further and behavioural aspects of demographic forecasting receive much less attention. Next, once working as a professional, the demographic forecaster talks a lot to other demographers and he becomes member of a group in which firm ideas about the demographic future remain unchallenged. For this reason Henshel (1982, 71) believes that US population projections made in the 1930s and 1940s came out far too high: demographers talked to each other too much. (It should be noted that this is consistent with the theory of rational expectations, applied to forecasters: when an individual forecaster formulates his assumptions regarding future demographic trends, he tries to make optimal use of relevant information, including the ideas of his colleagues. Thus the interaction between forecasters causes a regression to the mean future.)

But even if the forecaster were to work in isolation from other demogra-

phers, institutional aspects would affect his forecast. The forecast would be released not as his individual intellectual product, but as the statistical agency's view on future demographic trends. Hence two important questions arise: who approves the forecasts before they are released, and what consequences arise from the fact that it is usually a large statistical bureau that produces the forecasts, and not a small (academic) unit?
A questionnaire on national population forecasting in 30 industrialized countries in early 1988 showed that approval of the assumptions (and these are of major importance for the accuracy of population forecasts, much more so than the model) takes place within the immediate technical production unit in only eight of those 30 countries. In nine countries the assumptions are approved at an intermediate level within the agency, and in five countries this is done by the head of the agency. A further three countries have their assumptions approved by a government official above the agency's head, before the forecasts are released. In the remaining five countries, assumptions are reviewed and approved by an external group of demographers, planners and other professionals (in which the agency's forecasters are represented as well). Thus we see that in more than 70% of the countries the producers of the forecasts do not have the final responsibility for their product. This means that interests beyond the mere demographic may play a part in the majority of national demographic forecasts in industrialized countries.

Demographic forecasts produced by statistical agencies are characterized by authority and legitimacy. A number of factors contribute to this (see Keyfitz, 1984a, 27, 30; Ascher, 1978, 31-2). Firstly, such agencies are often impartial, and neutrality of a source contributes to legitimacy - whether of current data or forecasts. Policy-makers guarding their own preferences consider forecasts produced by a remote organization less threatening than "in-house" products. The agency's indifference to particular policies is an important feature of most official national demographic forecasts. It should be noted here that a forecast which includes policy intentions has to consider explicitly the effectiveness of policy measures with respect to demographic variables. In section 3.1.3 I discuss this issue further. Secondly, and more general than the first point, policy choices are rarely connected directly with specific population forecasts. (This argument does not hold, of course, with respect to policies aiming at birth control in order to reduce population growth, or with respect to policies on external migration - whether immigration or emigration). Thirdly, statistical agencies usually have better access to basic data than other potential forecasters, and this would make such an agency the only "natural" place for making demographic forecasts. This argument, given by Ascher, raises the issue of general availability of data. This availability seems to be less in the case of demography, than with economics. At least, pleas for statistical agencies to monopolise *economic* forecasting because of this data argument are seldom heard. On the other hand, the type of demographic data that are usually available to the statistical agency, leads more or less automatically to one specific forecasting methodology, viz. the cohort-component approach (see section 3.1.1).

The fact that national population forecasts are usually produced by statistical agencies may, generally speaking, be evaluated positively, considering the three reasons for authority and legitimacy. A drawback of this situation is, however, that the context in which those forecasts are made may lead to the introduction of forecast errors. Dennis (1987) investigates the impact of the decision-making context on forecast errors in the field of transportation planning. His conclusion holds equally well for demographic forecasts: in some decision-making contexts it is more important to be persuasive than to be right or wrong. Procedural rationality is a contextual feature that can significantly affect the generation and use of knowledge in bureaucratic settings. The timing of decisions is often based on pre-established deadlines, and truth is relative and negotiable. Given the fact that many demographic forecasts are scrutinized in a bureaucratic context, this introduces errors in those forecasts. Such errors are less likely to occur in non-bureaucratic settings.

3.1.3. The link between national demographic forecasts and political decision-making

Population forecasts at the national level are predominantly used by government planning organizations and other government agencies. The use by the private sector, the general public or the news media is much less frequent. This can be demonstrated by the results of the 1988 Survey on National Population Forecasts in Industrialized Countries, which was mentioned earlier. Statistical agencies were asked to rank the principal users of their forecasts. Six types of users could be distinguished: government planning agencies, other government agencies, the general public, business and industry, the news media, and consulting firms. The type of user who receives most information should be ranked as no.1, and no. 6 should be the one who is provided with the least information. No fewer than 27 of the 30 countries involved indicated that government planning agencies are the principal users of their forecasts, with an average score of 1.10. Other government agencies were ranked second in 23 countries, with an average score of 1.82. Business and industry, news media, consulting firms and the general public were ranked much lower as much less frequent users. Average scores for these four types of users range from 3.42 (business and industry) to 4.27 (general public).

Given this strong dominance of the public sector over other types of forecast users it is relevant to discuss the link between forecasting and political decision-making. What functions do demographic forecasts have for government planners and policy makers? To what extent do these users have an impact on essential features of the forecasts?

Forecasts may be used for a variety of purposes. In one extreme situation, they just *signal* future trends which are relevant for the user. At the other end of the spectrum we find a situation where they *legitimate* intended or planned policy actions. Intermediate situations arise when forecasts are used *to create an agenda of planning and/or research issues,* or when there is a strong *interaction between forecasting and planning.* The key question to distinguish between these purposes is whether the plan existed before the forecast was drawn up, or perhaps could have been developed without the forecast. When the answer is in the affirmative, the forecast was used as a tool for legitimation.
When forecasts in general are produced by an operating agency, the function of legitimation is usually more important than the warning function. However, as was noted in the previous section, decision makers belong most often to a different organization than population forecasters. Since in Western countries national population forecasts are rarely produced outside statistical agencies, most of these forecasts will be used as a warning instrument. Moreover, the group of users is fairly heterogeneous: planners in fields such as energy, regional policy, housing policy, education, labour market and many others. Hence, national population forecasts can be typified as *general purpose forecasts,* with the flow of information mainly directed from forecast producers to forecast users, and much less in the opposite direction. But this situation may very well change in the near future, as Baumgartner (1987) points out. Population may soon become a central policy concern. The socio-political consequences of declining and ageing populations in the industrialized countries, on the one hand, and of the growing and relatively young populations in Third World countries with unfavourable economic prospects, on the other, may well turn the population future into a political issue. Explicit birth policies and migration policies at the national level may be formulated (some Western countries already did so), directly aiming at bringing up or down numbers of births and numbers of migrants.

Should an independent forecaster translate such demographic policies into quantified assumptions on the relevant components? If he does so, given the definition of a forecast adopted in chapter 1, he estimates those policies to have most likely the effects they aim at in the future. This may be too strong an assumption. If he does not, he assumes more or less explicitly that these demographic policies will most probably not be effective. So we have a dilemma: Keyfitz (1984a, 35) states (and this holds for forecasting in general) that policy is the enemy of forecasting - and that is ironic since the purpose of forecasts is to help determine policy. Probably the best thing for our forecaster to do is to work as independently as possible, taking full responsibility for a forecast that does not necessarily follow explicitly formulated policies. He has to judge the effectiveness of these policies and to monitor closely relevant variables. Frequent updating may be necessary.

The situation sketched here is not purely theoretical, as is shown by the 1988 Survey on National Population Forecasts in Industrialized Countries. Statistical agencies in seven of the 30 participating countries stated that, when extrapolating fertility, they consider national plans or other official targets (together with other information, such as birth expectations, fertility theories and fertility trends in other countries). And for external migration this is the case in eight of the 24 countries that include external migration in their national forecasts. No general conclusion can, of course, be drawn as to *how* those policies are included in national forecasts, without a detailed analysis of the assumption-making process in each of the countries involved.

So far I have discussed the possible influence on population forecasts of one special group of users, viz. policy makers whose plans have a direct or an indirect impact on national population numbers. This group, of course, constitutes only a minority of all users. The majority show little interest in the production of a demographic forecast at the national level. They take the resulting numbers as they are. The reason is that, however great the uncertainty as to what the future population will be, the uncertainty in other variables is much greater. Uncertainties and errors of population forecasting may be lost in the larger errors of the other variables on which the plan depends - though this is, of course, no reason to be careless about the population part (Keyfitz, 1984a, 17).

Some users produce forecasts in their own field, using demographic forecasts as an input. However, the non-demographic developments may have a direct impact on the demographic ones. For instance, a labour market forecaster uses future numbers of females (and males) in working ages produced by the demographic forecaster. But his assumptions on the participation rates in the labour market of these females bear a direct consequence for the assumptions on childbearing in the demographic forecasts, since there is a simultaneous relationship between fertility and female labour force participation. This example demonstrates what Ascher (1978, 201) calls the interconnectedness of forecast tasks: the ability to forecast in one area depends on the ability to foresee how trends in other areas may impinge. Exogenous factors in one area are endogenous in another area, and vice versa. For instance, core assumptions in population forecasting are social, economic and technological; and those in economic forecasting are social, demographic and technological.

Since there is a division of labour between forecasters, we are faced with a problem here. At present, no one would advocate constructing one huge model which contains all factors. In the 1960s and early 1970s, the attitude towards large-scale models was much more positive than nowadays. It was the time of the world models of Jay Forrester, the Bachue model, and, somewhat later and smaller than these two models, the IMPACT model in Australia. But during the 1970s, the enthusiasm for large-scale models was dampened, partly because of their poor performance, and the high costs to develop and run them (Lee, 1973). Nowadays, a practical solution to the interconnectedness of forecasts is to coordinate assumption-making processes between forecasters in interrelated areas, and to make frequent updates of all those forecasts where links are strong. Or, more ambitiously, to formulate one overall scenario for the various areas, and next to derive sector-specific scenarios.

3.2 | Stages in the production of a population forecast

Within the broader context of social and political influences upon national demographic forecasting, I shall now isolate the mere technical issues of forecasting per se. This is what was called earlier the "social-engineering" perspective. The purpose of focusing on these technical issues is, again, to identify stages where subjectivity plays a role, and where sources of uncertainty may be present. The findings of this technical section should be seen in conjunction with those of section 3.1 on socio-political issues.

To make a quantitative forecast one needs a model, that is, a simplified abstraction of the real world. The model describes the behaviour of the system in question. Therefore, the *first stage* in the production process of a national population forecast is the *identification of the population system.* The population itself (the subject of the system) has to be defined, the population categories (the system elements) should be identified, as well as the relevant demographic relations between the system elements. Several points are important when a forecaster considers the characteristics by which the future population should be broken down (see also Brass, 1974, 537).

1 The purpose of the forecast and the user's requirements could indicate which population categories might be relevant. Because population forecasts are general purpose forecasts, the forecaster will be guided by the requirements of the principal user(s) only.
2 Theories explaining demographic behaviour (childbearing, migration) may reveal homogeneous population subgroups whose behaviour is relatively simple to interpret. This could ease the extrapolation of the future behaviour of these categories.
3 The available data may constrain the number and type of population categories to be distinguished.
4 Random fluctuations could occur when too detailed a classification were to be used, and cell-counts became too small.

In the *second stage* of making population forecasts the *description of the population system* takes place. The historical behaviour of the identified population categories is measured and analysed. If this behaviour is insufficiently understood, a revision of the identification stage may follow. For instance, one may find it necessary to include marital status in the population structure, or cause of death in mortality analysis. Demographic measurement concepts, time series methods, curve fitting techniques and multivariate models are the basic tools to be used.

The *third stage* consists of the *model construction.* A model is built which shows a one-to-one correspondence with the population system that was identified in the first stage. Sometimes one chooses a more simplified projection model (e.g. by assuming no interaction between mortality and migration).

The *fourth stage* of the population forecasting process, the *extrapolation* of the model parameter values is the most crucial one. For proper assumptions about the demographic future are more important for the success of a forecast than a sophisticated model (Ascher, 1978). Hence a systematic approach to the assumption-making process is necessary, and this may be facilitated by distinguishing several levels at which these assumptions could be formulated (see also Pittenger, 1977, 365).

1 Detailed assumptions apply to the future trends in age specific fertility rates, death probabilities by age and sex, etc.
2 Summarizing assumptions are formulated for summary indicators which aggregate the detailed parameters for fertility, mortality and migration over age (e.g. the total fertility rate, the life expectancy).
3 General assumptions are statements of a socio-demographic as well as a more general (social/ economic/ cultural/ political/ legal) nature. For instance, the assumption that the role of the family will become less dominant, or that legal regulations will have a limited impact on immigration. This level may be denoted as the "scenario level".
4 Disclaimers, which state that the forecasts are not valid in the event of war, natural disaster or major economic crises.

The discussion will be limited to levels 1-3.
In many situations, it may be useful to start assumption-making on level 3. Next, a qualitative assessment of the level 2 parameters may be given (rapid increase, stabilization). After quantification, the resulting summary indicators may be disaggregated into detailed parameters at level 1.
Different projection techniques may be employed for levels 1-3. The general assumptions may be drawn up by making use of socio-demographic surveys, scenario writing, Delphi methods or simply expert opinions. This also holds true for the qualitative assessment of the level 1 and the level 2 parameters. The numerical values of the detailed and summary indicators may be found by applying methods for extrapolation (when it concerns setting target values, see section 2.3.2) and interpolation (to bridge the gap between the most recent observations and the target values), which range from simple graphical extrapolations to advanced time series techniques or multivariate explanatory models.
Because of the inherent uncertainty in population forecasting, more than one set of parameter values is often made up for a particular component. In

section 2.3.1, two considerations were mentioned which may be used when choosing the phenomena for which to formulate more than one variant.

1 The significance of each individual component for the projection results. These results are more sensitive to changes in one component than to changes in other components. Identification of crucial component parameters requires sensitivity analysis.
2 The past course of the components. The more erratic the observed development of a given component, the more difficult it is to estimate its future course. Here, use is made of the results of the second stage, that of systems description.

The most crucial and/or most unstable parameters should be included in the variants of the forecast.

During the *fifth stage,* called *execution, documentation and implementation,* the projection model carries out the computations, the assumptions and results are recorded on tables, magnetic tapes etc. and they are published. Several factors may enhance or impede the implementation of the new forecast. Implementation is easiest when the forecaster and the user belong to the same organization. The other extreme is a situation in which an agency produces a general-purpose forecast, without much knowledge of its users, and in competition with other forecasting agencies.

The final phase of the production of a national population forecast is that of *monitoring.* Monitoring consists of a comparison between predicted and observed behaviour of the population system and an assessment of the deviations and their impact for the years to come. Hence, this last stage may also be viewed as a preparation for an updated version of the forecast. The results of the monitoring process may lead to changes in the model parameter values, but sometimes modifications of the projection model and extensions of the description of the population system are inevitable.

The various stages described above are summarized in figure 3.1.

Considered from the accuracy point of view, the three most crucial phases are: those of systems identification, systems description and parameter extrapolation. Subjectivity plays a major role in these three phases, and this introduces sources of forecast errors. Subjective decisions have to be made during the stage of systems identification, when the forecaster is confronted with the choice whether or not to include demographic characteristics such as region of residence, marital status, nationality etc. in the population. In the stage of systems description subjectivity is abundant, because here interpretation of historical trends and an assessment of the relevance of non-demographic variables for the explanation of these trends are required. And finally, parameter extrapolation is the most subjective of all forecasting tasks. It adds to the subjectivity of the previous tasks the fact that the fore-

Figure 3.1. Flow diagram for the construction of a population forecast

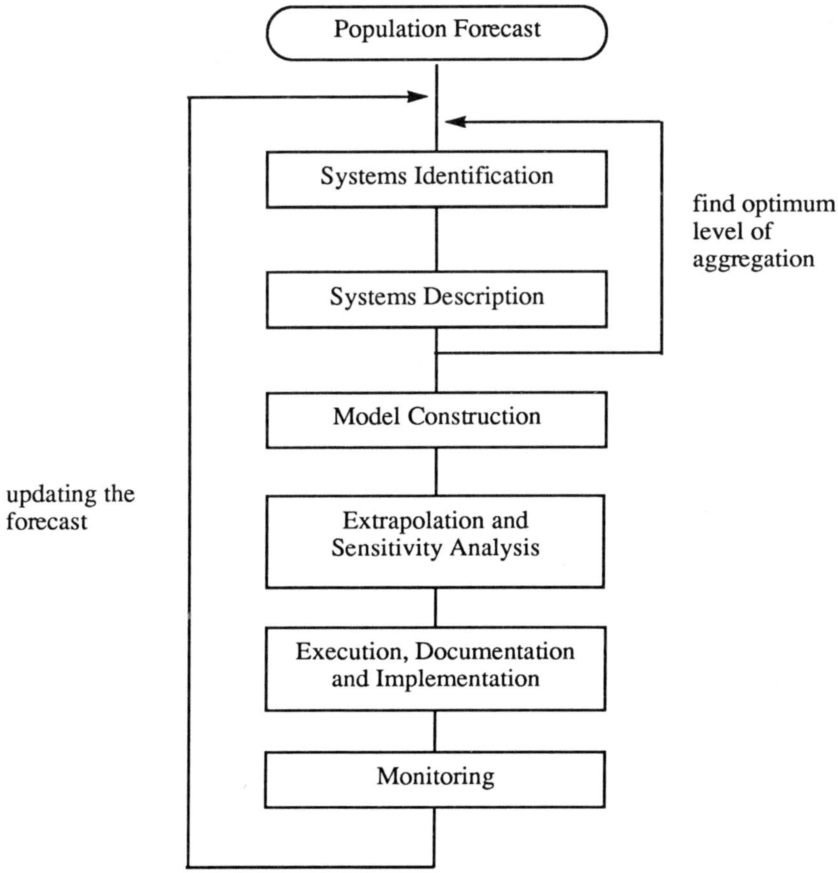

caster must now decide whether relationships assessed for the past, will also hold in the future. Quantitative models that explain demographic behaviour cannot be used here, for reasons to be discussed in section 3.3. But even if they were available, the choice for one particular model would be a subjective one - just as subjective as it would be to keep currently observed demographic rates constant, instead of opting for a smooth extrapolation of historical trends.

Theories which try to explain and predict demographic behaviour play an important role in the three stages of forecasting discussed above. Therefore in the next section I shall discuss the possible contribution of those theories to population forecasting.

3.3 | The contribution of theories to demographic forecasting

Ideally, population forecasts should be based on social theories linking social, cultural, psychological and economic factors to demographic behaviour. But in practice the core of population forecasting consists of extrapolation of observed series of fertility rates, mortality rates and migration variables, without a firm cognitive basis. This is because social science has, so far, not been able to formulate demographic behaviour in lawlike irrefutable and universally valid theories. In this section I shall discuss the reasons why social theories play a minor role in demographic forecasting, and what the implications are.

A number of writers have argued that Hempel's famous "covering law" model (also known as the deductive-nomological approach to explanation and prediction, see Van Seventer and Vossen, 1979, 33) is not the ideal form for all scientific thinking. The most important aspects of this model are:

1 science consists of a search for "general laws" to explain events occurring to objects;
2 the statement of a law makes a *generalization* across a domain of events or objects;
3 general laws connect events in patterns, referred to as "explanation" and "prediction".

Explanation and prediction have the same logical structure. *Explaining* an event E that has already happened consists of finding a suitable set of statements afterwards. These statements consist of two groups: antecedent conditions and general laws. We speak of *predictions* when the antecent conditions and the general laws are known and we derive E prior to its occurrence.
The pursuit of general laws is characteristic primarily of the physical sciences. Other sciences have been much less successful. D'Andrade (1986) distinguishes two other groups of sciences besides the physical sciences. Firstly, the natural sciences (biology, geology, some aspects of meteorology and oceanography, much of economics and psychology, and some fields of anthropology and sociology) have found some generalizations of "how things work": DNA, the cell, the heart, the body, the ecological system. But these generalizations are much more limited than those in the physical sciences. ".... these kinds of quantitative statements are actually the derived outcomes of other more basic processes, rather than basic laws in themselves" (D'Andrade, 1986, 21). Secondly, what he calls semiotic sciences (linguistics, some fields of psychology, anthropology, and sociology) consist of those fields that study "imposed" order based on "meaning" rather than on natural or physical order. As imposed order that creates meaning D'Andrade mentions the normative uniformities of dress and behaviour, material culture, a dance or ritual, nation-states, statuses, laws. What the semiotic sciences seek to answer are questions about how order is generated, or what kind of more abstract order lies behind the order we see (D'Andrade, 1986, 23).

Both the natural science approach and the semiotic science approach make generalizations, but the generalizations are quite different in character. The generalizations of natural science sometimes sound like laws (except for the long list of often unstated qualifications). But semiotic-science generalizations typically sound like interpretations, totally contingent on time, place, and person. What the two approaches have in common, is that the covering law model fits them badly (although the fit is worse for the semiotic sciences than for the natural sciences). Indeed, Boudon (1986) argues that definite "laws" are in principle beyond the reach of theories of social change, and Henry (1987, 10) supports this view in the case of demography. Hence extrapolation to new circumstances apparently has to rely on *qualitative* beliefs about the processes at work, and forecasts depend on substantive conjectures - not on law-like generalizations framed in quantitative statements. The latter point is neatly demonstrated by Keyfitz (1982) for the case of demography.

Keyfitz attempts to discover whether theories have improved demographic forecasting. He analyses possible contributions from insights concerning among others the fertility transition, effects of development, the Caldwell theories concerning education and fertility, urbanization, income distribution, Malthus' writings on population, human capital, the Easterlin effect, opportunity costs, prosperity and fertility, the search for homogeneous subgroups and childbearing intentions. Many items in this list could not be termed theories, in the sense that they describe universally (irrespective of time and place) valid relationships - yet they represent the core body of substantive knowledge in demography.
Keyfitz's conclusion is a pessimistic one: many important theoretical relations derived from perfectly valid research cannot be usefully incorporated in forecasts. He gives a number of reasons. First, much theory is neither intended nor usable for prediction, since that theory is cast in the form of comparative statics and provides conditional results. Prediction, however, requires dynamic and unconditional models. Second, the theory may be true but not general enough to be robust against other relations not contained in that theory. Third, the theory may be true, but it may be imprecise as to the timing of the events it predicts. Demographers argue in terms of quantum and tempo. Hence they are not satisfied when they know *that* a certain event will happen (quantum) without being able to say when, at which age (tempo). Fourth, the theory may involve exogenous factors which themselves are difficult to foresee. Fifth, the relation between explanatory variables and the variables to be explained can change unpredictably. Sixth, everything about the relation may be known, but of too short term for the forecast intervals that are of great interest.

Keyfitz's findings are of an *empirical* nature: he noted that demographic theories have proved to be of only limited value to demographic forecasters. However, a finite list of failures does not necessarily imply that we cannot formulate true laws. Therefore, it is useful to summarize Boudon's argument, which is *theoretically* underpinned.

Boudon's arguments apply to theories of social change, and I assume that they apply to demographic behaviour as well, in particular to childbearing and union formation and dissolution. The dramatic changes which took place in the Western World since the sixties in these two demographic components (and which went together with socio-cultural and socio-economic processes like individualization and emancipation) may very well be typified as major social changes.
Boudon (1986, 32-45) discusses individual action and aggregation effects in terms of an analytical framework. A key feature of this model is that social change is to be analysed as the result of a set of individual actions. A phenomenon M is to be explained. M is a function $M(m_i)$ of a range of individual actions m_i for various persons i (i = 1, 2, ...). The mapping from m_i to M is called *aggregation.* This takes sometimes the simple form of addition (mortality as the sum of individual deaths), but more often the form of M is much more complex (for example, in the case of mortality due to infectuous diseases, the interaction between individuals has to be taken into account). Each individual action m_i is a function $m_i(S_i)$ of the structure S_i of the situation including the actor i. In turn, the structure S_i is a function $S_i(M'_j)$ of factors M'_j (j = 1,2,...) at the macrosocial level, the level of the system in which M occurs[1]. Explaining M is saying exactly what the functions M, mi and S_i are. Psychologists elaborate $m_i(S_i)$, or $m_i(S_i(M'_j))$, but the question "why M?" cannot be answered fully at the individual level of m. The macrolevel at which M occurs should also be considered, and aggregation should be checked. Explanation of the structure and comprehension of the individual actions are inseparable aspects of the analysis (Boudon, 1986, 38). Since microsociological statements m are determined by the nature of the question "why M?" there can be no *general* microsociological model, Boudon argues. In our case that implies that the demographer constructs *distinct* models for childbearing, marriage, the formation of consensual union, home-leaving, mortality etc.
Boudon discusses *conditional* laws (if A, then B) within the analytical framework sketched here. Such a law can symbolically be expressed as $B = M(m_i(S_i(M'_j;A)))$, or $B = MmS(M';A)$ for short. Suppose we have established the conditions K, under which, when fully met, we are certain to observe the 'if A then B' relationship. K creates an *ideal* situation, and only in those very limited K-situations we can be sure of seeing the relationship. Therefore, as Boudon concludes (p. 65) the demographer, together with the sociologist and the economist finds himself in a situation which is very different from that of the physicist. The laws in social sciences hold good in certain limited areas defined by K only. These areas have almost always vague boundaries and it is in fact impossible to describe K exhaustively. To summarize ".... the limited areas in which laws hold good, the vagueness of

the boundaries of those areas and the difficuly of determining what the real circumstances of a concrete situation might be are the three factors that play a part in giving the notion of constitutional (sic, NWK) law a different epistemological status in the natural and the social sciences." (Boudon, 1986, 66)[2].
Therefore, Boudon makes a sharp distinction between universal laws, which apply to all cases, and local laws, which apply to particular cases only. Hence, relationships should be formulated existentially ('if A, then sometimes B') and never deterministically ('if A, then always B'), or probabilistically ('if A, then most frequently B'). The most we can say is that A *may* be followed by B. Such laws are expressing mere possibilities.

The point that generalizations in the field of social change are narrow in scope is not new. Shweder and Fiske (1986, 4) remind us of the fact that Nagel made the same point for social science in general in 1961 in his book "The Structure of Science", thereby citing a reference from 1934. They cite Nagel who wrote in 1961 that ".... (un)like the laws of physics and chemistry, generalizations in the social sciences... have at best only a severely restricted scope, limited to social phenomena occurring during a relatively brief historical epoch within special institutional settings." In other words, a nontrivial law can be universal only for a homogeneous set of objects or events.
Perhaps certain regularities will be established, but this requires at least the classical proviso of 'identical conditions'. That can only happen in particular cases, and hence the search for conditional regularities cannot be a major object of the analysis of social change.
Some social scientists believe that the postulate of determinism is an indispensible foundation of scientific knowledge. However, there are such things as incompletely foreseeable innovations (and this was perhaps what occurred in demographic behaviour in the Western World during the 1960s, see De Feijter (1989) for the Netherlands). But a more important objection against this postulate is that it requires that all social processes be closed, i.e. that it should be possible to establish the state at time t+1 of these processes from their state at t. At the other hand, open processes can behave unpredictably. Whether a process is open or closed is determined by the nature of relationships between the constituent elements of the process, and the inability to predict is a consequence of the structure of an open process. A closed process arises when there is a reduced area of choice, and a lack of any action by the actors in that area. However, processes of this kind are the exception rather than the rule. Most processes are open and the logic of the situation defines a range of options without offering any a priori convincing reasons for choosing one rather than another (Boudon, 1986, 165). The conclusions are clear: first, the existence of open processes is one major reason for believing that there are no good grounds for accepting a deterministic representation of social change; second, action does not necessarily take the form of a choice between two predetermined actions - it can also be inno-

vative. Boudon (1986, 173 ff.) discusses the concept of chance, and the part it plays in theories of social change: "... most (social scientists) think of it as merely the result of our ignorance, holding that we only see a phenomenon as not being fully determined because we do not have access to all the variables that shape it ...". Because regularities in complex situations are seldom, if ever, fully known, these scientists think they can formulate only incomplete propositions reflecting those regularities, from which inferences can be made *with probability* about the relation between a cause and its effect. "... Although such causes are not sufficient for their effects, the causing itself is not necessarily probabilistic..." (Marini and Singer, 1988, 357). According to this view, the causal relationship is deterministic, and probability comes in because not everything is fully known. However, Boudon argues that chance must be seen as a structure which *is* characteristic of certain causal relationships. Whether it is present or absent depends on the process in question.

The implications of this analysis and these conclusions are far-reaching. Science sets out to answer questions about reality, and there are three types of such questions. First, the form of the question determines whether the validity of its answer can be verified. Questions of this type are scientific questions. Second, the validity of the answer to the question cannot be established, but that answer is useful and plausible. In this case we speak of conjectures. Third, the question itself may produce irreconcilable answers, in which case we are dealing with metaphysics.
Most of the 'laws' formulated by theorists of social change are in fact conjectures. They are statements about what may happen. Sometimes a conjecture is so convincing that we can talk of a conditional law of the type 'if A, then B'. Hence there is a logical progression from statements of possibility via conjectures to conditional laws (Boudon, 1986, 191). To formulate a law (perhaps even a conditional one) requires two conditions to be met. First, a number of factors produce a readily foreseeable change in the motivation of some actors. Second, this change will in fact modify the actor's behaviour.
Thus many theories of social change are, strictly speaking, unacceptable, because they conflict with many of the observed factors. They should be considered as formal constructions that are useful because they isolate an ideal mechanism that cannot be oberved in a pure state in reality. It is only possible to construct theories of social change about partial and local social processes firmly situated in time and particular circumstances. In other words, the only scientific theories of social change are those that are *partial* and *local* (Boudon, 1986, 208). General theories of social change do not exist.

Given Boudon's profound analysis of theories of social change, it is not difficult to understand why Keyfitz was forced to conclude that demographic theories do not really contribute to population forecasting. First of all, many 'theories' that Keyfitz investigated are formulated at the individual level (they contain statements of the kind mS(M')), but they are tested at the macro-level of M. Hence, they confound individual behaviour and aggregation effects. Second, and more important, the degree of closedness of these processes is so low that the state of the process at time t+1 cannot be derived uniquely from the state at time t. Exogenous factors play their part, which leads to open processes that can only be described by a set of conjectures, or at most a set of conditional laws. When we call this a theory we should be aware that it is partial and only locally valid. What we *have* consists, strictly speaking, of two concepts of knowledge: an *empiricist* concept, and a *nomological* one. With the first type of knowledge I intend to describe developments 'wie sie eigentlich geschehen sind' (Boudon, 1986, 223). The second type of knowledge involves established trends, regularities, cycles, laws governing which phenomena follow others, covariations between series, causual links between factors etc. What we *seek* is knowledge according to the *formal* concept: hypotheses are used to deduce consequences, and we employ the model thus created to explain, illustrate or understand reality. Next, this model is used to predict demographic behaviour.

Boudon's analysis illustrates that definite laws in the sense of Hempel are beyond the reach of theories of social change, and hence, I add, beyond the reach of theories on demographic change and demographic phenomena. The empiricist and nomological knowledge that is produced in demography fits perfectly well in D'Andrade's second and third group of sciences, viz. the natural sciences and the semiotic sciences. Hence demography consists of conjectures, not of laws. This means that forecasts based on demographic knowledge will time and again be falsified by reality. Indeed Brass (1985, 41) believes there will be advances in the grasp of the social determinants of fertility, and hence birth projections will become more reliable. But not if it demands the forecasting of social change. I add Kingsley Davis' (1979, 48) observations on the blind character of contemporary social change, caused by the continuing process of individualization: "Given the character of human societies as collections of individuals, the unplanned and unforeseen character of social change seems inevitable". Vossen (1988, 41) gives essentially the same argument for his pessimistic view on potential contributions of social science theories to demographic forecasting.
Does this imply that demographic forecasting is impossible? This conclusion is too pessimistic, in my view, for two reasons. Firstly, most of the discussions in the previous sections relate to what I called *inherent uncertainty,* see section 1.3.2. But in practice one may hope to encounter in a given situation milder forms of uncertainty, too: randomness and ignorance. In that case one will be able to formulate fragments of (local and partial) laws. Secondly, demographic forecasting does not presuppose the use of demographic theories, see the definition of forecasting in section 1.3.1. It would be of great help, no doubt, to have a valid theory, but it is not absolutely

necessary. As a second best, the population forecaster extrapolates observed regularities in time series, supplemented by fragmentary information (cf. above) obtained from such diverse sources as fertility surveys, expert opinion, trends in 'forerunner countries', and so on. Fortunately, demographic structures generally evolve only slowly - the population system shows a great deal of inertia. But the other side of the coin is that, exactly because of this approach to demographic forecasting, trend shifts often cannot be foreseen adequately. Even, demographic forecasters frequently lag behind actual developments in times of instability.

It should be stressed that I do not believe that psychologists and social scientists who are not able to predict individual behaviour do not have a good science. A serious, but vain attempt to discover regularity does not disqualify the scientist. Prediction and explanation are very difficult because particular events have idiosyncratic boundary conditions, as I have just argued. Hence whether a science has its merits should not be judged by its ability to predict or explain particular events (D'Andrade, 1986, 28). Description and partial understanding of reality is a perfectly legitimate goal for any social science. Moreover, attempts to solve the unsolvable have their merits: the mathematical problem of how to trisect an angle with ruler and compass resulted in group theory. Thus, as Keyfitz (1985a, 60) argued when he discussed the problematic nature of demographic forecasting, the preoccupation with a hard problem challenges ingenuity and expands the intellectual territory of a discipline. Although it is difficult to live with social science predictions, at the same time it is impossible to live without them (Henshel, 1982, 59). Several reasons in favour of social science predictions can be given. First, social science, like any science, has the foremost duty of producing predictions. Second, public forecasting of social matters will be attempted anyhow, because of the apparent need of planners and policy makers for predictions. Third, the task of prediction imposes a certain degree of rigour on thinking (Henshel, 1982, 58).

3.4 | Hypotheses on accuracy and errors in national demographic forecasts

The overall aim of this chapter is to contribute to the formulation of a theory of forecast errors in the framework of national demographic forecasts. Why and how do such errors occur? Which factors cause their amplitude, and their sign (over- or underestimation)? How do they develop with time?
It will be clear that such a theory of forecast errors will not contain definite laws in the sense of Hempel. There is a strong parallel between Boudon's analysis of social change processes and processes that generate forecast errors: both processes are predominantly open, that is, there are no compelling reasons why the actors in such processes (individuals in the case of social change processes, variables expressing the deviation between model parameters and corresponding observed parameters in the case of error pro-

cesses) should display the behaviour that we observe. For the case of social change this is discussed at length by Boudon. That error processes are open can be seen as follows.

Take a recent population forecast, and consider as one of its results the number of births expected for next year. The error in that number is generated by an entirely open process, since, given the forecast, we can do nothing but to wait and see what will happen to actual births in the course of next year. And as these births are generated by an open process (as argued in the previous section), the error in the number of births is generated by an open process, too.
Hence the theory of forecast errors to which I seek to contribute will consist of a set of conjectures, similarly to theories of demographic behaviour. Moreover, it will contain generalizations describing "how things work", rather than laws in the sense of Hempel. Therefore, it will be much closer to theories in the natural sciences, than to those in the physical sciences (see D'Andrade's typology of sciences discussed in section 3.3.1). The regularities I possibly will achieve are a set of more or less general trends, and some conditional "laws" of the form 'if A, then (usually) B' (probabilistic).

In what follows I discuss a number of factors which one may assume are related to the accuracy of population forecasts. Some of the considerations follow more or less immediately from the theoretical arguments given in section 3.3. Others are based on empirical findings in the literature on forecasting. Many factors which are (assumed to be) associated with forecast errors have been mentioned. The review below is limited to factors which are relevant to the existing practice of population forecasting at the national level. Therefore I do not discuss wider issues involved in the analysis of accuracy-related factors in forecasting, such as objective versus subjective methods, naïve versus causal methods and linear versus classification methods. Hypotheses on which methods will be most accurate in which situations are extensively discussed and tested by Armstrong (1985, chapter 15). The issues which I found relevant to national population forecasting are grouped into three broad classes: time-specific factors, forecast-specific factors and miscellaneous factors.

* *Time-specific factors*

1 Since demographic behaviour cannot be explained in the causal sense, demographers construct their forecasts on the basis of observed regularities. Hence, demographic forecasting may be interpreted as the search for population functions which are more or less constant through time (Keyfitz, 1972, 347). Trend extrapolation of observed regularities, taking place in the fourth stage of a forecast's production (see section 3.2) plays a key role in demographic forecasting. Therefore, a first hypothesis is that *irregularities will tend to increase forecast errors.* These irregularities may take on several forms.

1.a Firstly, Ahlburg (1982) notes that the Bureau of the Census underestimated births in the upswing and overestimated them in the downswing. This phenomenon may be interpreted as one manifestation of what Ascher (1978, 53) calls "assumption drag": the continued use of assumptions long after their validity has been contradicted by the data. In both cases, a *steep increase or decrease* of observed developments would tend to increase forecast errors.

1.b Secondly, a possible source of errors is *fluctuations in observed data,* which may be assumed to go together with decreased accuracy. Makridakis and Hibon (1979) find a strong impact of this factor on the accuracy of economic time series, and Ahlburg (1982, 372) demonstrates that errors in births forecasts correspond reasonably well with the degree of instability of birth expectations, as recorded in fertility surveys. On the other hand, Ascher (1978, 51) notes a surprisingly small influence of the volatility of ex-post observed population size on errors of forecasts of total population in the United States between 1938 and 1949. Also Henshel (1982, 74) argues that demographic forecasting is atypical in this respect, compared with other forms of social forecasting.

1.c Finally, *current levels* of key demographic indicators dictate to a great extent their forecasted future levels when subjective extrapolation methods are used. Long (1987, 11) notes a strong correlation between total fertility rates in the base years in which post-World War II forecasts were made in the US, and the projected ultimate fertility rates in the target years. In a recent survey of social forecasting in the Netherlands for the period 1977-1987, it was noted that expectations are strongly influenced by the situation at the time a forecast is made, because social dynamics are difficult to appraise at the time when they take place (WRR, 46). Thus, target values for year t+n are closer to observed values in the jump-off year t, than to actual values n years later, i.e. in year t+n. This phenomenon may be understood as another manifestation of the assumption drag described above.

2 Given a certain forecast, the influence of *ex-post observed population trends* on forecast accuracy is important by definition. The relevance of observed trends for errors of population forecasts is stressed by Ascher (1978, 51), by Page (1982, 451), by Long and McMillen (1987, 150) and by Long (1987, 10) for the United States and by Bretz (1986, 255) for the FRG. Henshel (1982, 74) does the same for the wider field of social forecasting. In general, the finding is that relatively accurate demographic forecasts are made in times of smooth trends, but that erratic demographic developments do not sufficiently account for inaccurate forecasts.

* *Forecast-specific factors*

3 Stoto (1983) shows that *forecast's jump-off year* is strongly correlated with the errors in population growth for six forecasts made by the US Bureau of the Census between 1946 and 1971 and covering the period

1950-1975. Pflaumer (1988b, 115) states that the impact of the jump-off year even dominates that related to the length of the forecast period. However, neither authors disentangle properly forecast effects, duration effects and period effects. In chapter 6 I separate these effects, and I analyse errors in fertility and mortality.

The jump-off year of the forecast cannot be an explanatory factor for forecast errors in the causal sense. It is just a label for all factors that are unique for the particular forecast. A sharper hypothesis than the previous one is the following.

4 A *change in forecast approach* will not be introduced if this were to deteriorate forecast quality. Examples of such changes for industrialized countries are the increased interest in longitudinal analyses and extrapolations of fertility since the 1960s, the introduction of cause of death in the component of mortality, and the introduction of the component of external migration in the forecasts (of the 30 countries which took part in the 1988 Survey on National Population Forecasts in Industrialized Countries, seven declared that they do not include external migration in their official national forecasts, see Cruijsen and Keilman, 1989). Forecast approach is not confined here to aspects of the demographic model or to formal characteristics of the extrapolation procedure. I also include issues of data collection. For instance, childbearing intentions from fertility surveys may be used to increase forecast quality. More generally, fertility surveys provide information on sociodemographic, socioeconomic and sociocultural backgrounds of childbearing behaviour. This information may be incorporated in the forecasting process. Therefore, the hypothesis is that changes in forecast approach in the sense just described will tend to decrease forecast errors.

Changes in forecast approach are studied by Ahlburg (1982). He analyses the consequences of the introduction of a longitudinal perspective in 1964 in forecasts of births in the United States. He finds some evidence for an improvement of forecast accuracy connected with the move from a period methodology to a cohort methodology. In chapter 6 I argue that his conclusions may be biased because of confounding period and duration effects.

It should be noted that simple forecasting methods may perform more accurately than, or at least as accurately as, sophisticated methods. This was found in a number of studies - for reviews see Mahmoud (1984) and Armstrong (1985). Stoto (1983) demonstrates that for projections of total population size, simple projection techniques are more accurate than more complex methods.

5 The issue of whether or not to use *experts,* and its impact on the accuracy of three time-series methods, was investigated by Carbone et al. (1983) - see also Mahmoud (1984). The conclusion is that technical expertise, judgemental adjustment and individualized analyses were of little value in improving forecast accuracy as compared to black box approaches, which can be applied rather mechanically. Alho and

Spencer (1988) suggest that expert opinion may have caused a systematic bias in US mortality forecasts.

6 Somewhat surprisingly, I was able to find only one empirical study into the impact of the *organizational context* on forecast errors. In section 3.1.2 I mentioned the analysis of Dennis (1987) in the field of transportation planning. His findings, when applied to demographic forecasts, lead to the assumption that errors are introduced in those forecasts because of procedural rationality in bureaucratic settings. Hence, any decrease of the decision-making power of such a bureaucracy may be assumed to increase forecast accuracy.

7 Makridakis and Hibon (1979) use 111 time series to examine the accuracy of various time-series methods. They investigated the Mean Absolute Percentage Error (MAPE) of model fitting and that of forecasts. The *number of data points* in the series have a limited impact on the MAPE of model fitting - the influence of this variable on the MAPE of forecasting was either nonsignificant statistically, or when it was significant, it was of the wrong sign. It would be natural to include this factor in the analysis to account for the length of the base period for extrapolation. The assumption, then, is that too short or too long base periods produce inaccurate forecasts.

8,9 Makridakis (1984), in his introduction to the book on the forecasting competition which followed the 1979 article by himself and Hibon, concludes that (8) the *type of data used* (yearly, quarterly, monthly, micro, macro) and (9) the *length of forecast period* (also denoted as *time horizon of forecasting* or *forecast duration*) affect forecast accuracy. The decrease of accuracy with increasing length of forecast period is discussed, among others, by Ascher (1978), Stoto (1983), Armstrong (1985) and Smith (1987).

10 Empirical evidence suggests that the rule "once an over (under) estimation, always an over (under) estimation" holds true in a fairly large number of cases (see for instance Keyfitz, 1981, 589). What counts is the direction of the initial take-off. Only in a few cases does the forecast curve cross the realized curve. This means that, for a certain forecast, *the error observed one year earlier* would be a good predictor for the present error. One would expect that this rule works better for fertility than for mortality or migration. The reason is that forecasters extrapolate trends, without any random fluctuations around these trends. Now, the trend in fertility is usually more important than random fluctuations, but this is not the case for mortality and certainly not for migration (see also the discussion below on the issue of frequency of measurement).

11 Various studies have indicated that a *combination of forecasts* often produces smaller errors than each of the constituent forecasts, see for instance Bunn (1988), Armstrong (1985, 183-186), Mahmoud (1984), and Newbold and Granger (1974). This can be made clear intuitively by pointing out that no single forecasting method is the best one in a variety of situations.

* *Miscellaneous factors*

12 An interesting hypothesis was put forward by Bartlema (1987, 21). Forecast errors are highest where behaviourally determined *demographic variables* are concerned: migration, nuptiality, fertility. In the terminology of Boudon (1986) this would be explained by the fact that the choices which individuals make, lead to open, and hence unpredictable, processes. Mortality would lead to relatively small forecast errors. The behavioural aspect is one argument which supports this hypothesis. A second argument is to be found in analytical demography. Migration, nuptiality and fertility are demographic phenomena which can be described by two aspects: their level (quantum) and their timing (tempo). How often does an individual migrate? How many children does he or she have? These questions are related to the *level* of the phenomena in question. Their *timing* is expressed by the age at which the corresponding events take place (or, more general, by the duration since some initial event). But for mortality, only questions related to timing are relevant. Since everyone dies, its level is 100% by definition. This makes mortality easier to foresee than fertility, nuptiality and the like.

13 Some authors have noted a negative correlation between *population size* and forecast errors. Indeed, regional forecasts show comparatively large errors as internal migration is relatively difficult to foresee (Schéele, 1981, 7; Brunborg, 1984, 96; Smith, 1984, 11; Birg and Brock, 1986, 33). But even without migration, the law of large numbers tells us that variances will be large.

14 Although population size may have a dampening effect on forecast errors, *population growth* tends to increase those errors. This has been observed at the national level (Keyfitz, 1981, 584) as well as on the subnational level (Stoto and Schrier, 1981, 28; for a summary, see Pflaumer, 1988b, 117). More generally, at the latter level the relationship even seems to hold for *absolute* growth, irrespective of sign: Smith (1984, 11) observes that counties in the US with negative growth rates and those with positive growth rates have relatively large forecast errors of population size. This finding is in line with the impact of the slope of current developments on accuracy which was discussed in the section on time-specific factors.

15 The *frequency of measurement* (length of unit forecast interval) may be related to forecast accuracy. Siegel (1972, 66) states that cyclical fluctuations cannot be projected as projections are relatively smooth. De Beer (1988, 4) notes that immigration developments for the Netherlands consist of a cyclical pattern with a period of approximately six years, superimposed on a long-term steady trend. Therefore, annual immigration forecasts tend to display relatively large errors, compared with a forecast of the average immigration level over a six-year period. The same phenomenon can be noted for the life expectancy at birth in the Netherlands, see figure 5.7 to be discussed later. However, the amplitude of the cycles is much less pronounced here than in the case of immigration.

16 *Technological change* may be assumed to have an impact on forecast errors. This factor summarizes various autonomous forces, such as the increasing body of scientific knowledge and of available data, continuous developments in computer hardware and software, intensified international cooperation, etc. The hypothesis is that with technological progress, forecast errors become smaller.

NOTES

1) Boudon (1986, 30) defines a function $S_i(M')$ of just one factor M', which, in my interpretation, is a vector with elements M'_j.
2) Anton Kuijsten pointed out to me that the word 'constitutional' in this sentence is probably an error arising from the translation of the original French book. One should read here 'conditional', of course.

4. TRENDS IN POPULATION FORECASTING METHODOLOGY IN THE NETHERLANDS, 1950-1987

The aim of this chapter is to describe the methodology of the national population forecasts produced in the Netherlands since 1950. Major methodological shifts will be identified. Thus the insights obtained here will serve as an input to the empirical chapters 5-7, where forecast errors since 1950 are analysed. In view of the disrupted social trends during World War II and in the early postwar years, the 1950s are a natural starting point. Sources (all in Dutch) are:

- NCBS (1951), which contains a review of national population forecasts of the Netherlands that were produced between 1921 and 1951;
- Smulders and Verhoef (1973) who discuss forecasts made by the Netherlands Central Bureau of Statistics (NCBS) between 1942 and 1971;
- Keilman (1982) who compares NCBS-forecasts of the 1960s with those made between 1975 and 1980;
- Vossen (1988) who briefly describes a number of aspects (societal and institutional context, forecasting methodology and approach of formulation of assumptions) of all published national forecasts of the Netherlands' population since 1921; and
- Van de Mortel (1988) who explores methodological aspects of national demographic forecasts prepared for the Netherlands since 1945.

The overview starts with the 1950 NCBS-forecast, i.e. the forecast which took the observed population of 1st January 1950 as its jump-off population (NCBS, 1951). Table 4.1 presents the demographic characteristics used in the NCBS forecasts since 1950, their components, the way in which fertility, being the major component of change, is modelled, the manner in which assumptions are formulated, and the number of variants which can be distinguished for each forecast. Several trends can be noted.

Table 4.1 Trends in population forecasting methodology since 1950, the Netherlands

	Jump-off year of forecast									
	1950	'51	'56	'65	'67	'70	'72	'75	'80	'84-'87[5]
Characteristics										
Age	x	x	x	x	x	x	x	x	x	x
Sex	x	x	x	x	x	x	x	x	x	x
Marital status M									x	x
F						x	x	x	x	x
Components										
Birth	x	x	x	x	x	x	x	x	x	x
Death	x	x	x	x	x	x	x	x	x	x
External migration	x[2]	x[2]	x[2]	x[2]				x	x	x
Marriage M									x	x
F						x	x	x	x	x
Modelling birth										
Total										x
Marital	x	x	x	x	x	x	x	x	x	
Extramarital	x	x	x	x	x	x	x	x	x	
Age-specific	x	x	x	x	x	x[3]	x[3]		x	x
Marriage duration-specific						x	x	x		
Birth order-specific								x	x	x
Assumptions										
Birth	p/1	p/1	p/1	p/1	p/1	c/2	c/2	c/3	c,p/3	c,p/3
Death	p/2	p/2	p/2	p/2	p/2	p/2	p/2	p/2	p/3	p/3
External migration	p/1	p/1	p/1	p/1				p,c/3[4]	p,c/3[4]	p/3
First marriage						c/2	c/2	c/3	c,p/3	c,p/3
Remarriage						p/1	p/1	p/1	p,c/3	p,c/3
Divorce						p/1	p/1	p/1	p,c/3	p,c/3
Variants							H,L	H,L	H,M,L	H,M,L

Notes:
1. First code: p = by period; c= by cohort; c,p = by cohort and by period, with emphasis on cohort. Second code: 1= detailed assumptions; 2= summarizing assumptions; 3 = general assumptions.
2. Basic variant without migration, one (1950) or two (1951, 1956, 1965) extra variants with net migration deficit.
3. Marriage duration-specific fertility, broken down by age at marriage.
4. Cohort-return migration patterns; other migration parameters; by period.
5. Forecasts with jump-off years 1984, 1985, 1986 and 1987.

4.1 | More components and more assumptions

First of all, the number of demographic components used in the NCBS-forecasts has increased. With the introduction in 1970 of female nuptiality, and in the 1980-forecast of male nuptiality, the most recent forecasts draw up assumptions for 7 demographic phenomena; in 1950 only 2 demographic phenomena were included.
Secondly, the level at which the assumptions are formulated is gradually rising. Table 4.1 distinguishes the same levels of assumption-making as those mentioned in section 3.2.
Until the forecast of 1967, *fertility* was described at the most detailed level, that of age-specific marital fertility rates. From 1970, summarizing indicators, such as the average number of children per woman, have been used. The year 1975 marked the first time that the NCBS formulated fertility assumptions at the general level: future fertility trends were viewed from a general and social-science point of view, in which the demographic, psychological, cultural and socio-economic factors influencing these trends were taken into account. In doing so, the NCBS supported the findings of the then State Commission for Population Issues which formulated concrete expectations for future fertility, mortality and nuptiality trends.
For *mortality,* age-specific rates were aggregated in the life expectancy at birth as early as 1950, using the life table method. Since 1980, the influence on mortality of preventive medicine, hygiene, nutrition and socio-economic, cultural and technological trends has been taken into account.

4.2 | Period or cohort, and age or marriage duration in fertility extrapolations?

Thirdly, the cohort approach to formulating assumptions is becoming increasingly popular. Simultaneously, application of the period approach is losing importance. In the entire post-war period, mortality assumptions have been drawn up by period analysis. However, for fertility there was a change from a period to a cohort approach between 1967 and 1970. Also, female first marriages were extrapolated by birth cohort for the first time in the 1970-based forecasts. This trend continued when in 1979 experimental calculations for marriage for single males applying a cohort perspective were published (NCBS, 1979, 18). Male nuptiality was integrated in the 1980-based forecasts, in which also some attention was given to a longitudinal treatment of divorce and remarriage (NCBS, 1982, 38).

The change for fertility from a period approach to a cohort approach, which took place between 1967 and 1970, was stimulated in particular by the Working Group on Population Forecasting Methodology. This Working Group was instituted in 1967, and included, besides the NCBS-forecasters, staff of the National Physical Planning Agency and of the Central Planning

Bureau, as well as some prominent academic demographers. Prompted by the poor results of the 1965-forecast (to which I shall return in chapter 5), this working group examined the advantages and drawbacks of a period methodology as compared to a cohort approach to the analysis and extrapolation of fertility. The advice given, and incorporated in the 1970-forecast, was two-fold: do not follow the usual period approach, where the results are influenced in particular by short-term effects. Instead, apply a cohort analysis which produces a better picture of the actual, more structural trends. And do not use birth cohorts (and age-specific fertility trends), but instead use marriage cohorts (and a marriage duration-specific analysis). The introduction in 1970 of marriage duration-specific fertility, from a cohort point of view, may be considered to be one of the most important methodological changes in national population forecasting methods in the Netherlands since the Second World War.

The change from a period perspective to a cohort perspective is understandable. During the 1960s, Ryder's insights from translation theory had gained acceptance, and U.S.-forecasts witnessed such a change in 1964. What is striking, however, is the emphasis on marriage duration, instead of age. Which reasons were given by Dutch forecasters for the adoption of marriage duration in the analysis and extrapolation of fertility? The arguments mentioned in the NCBS-publication of the 1970 forecasts are very general. First it is noted that demographic trends in the Netherlands during the 1960s are mainly characterized by changing patterns of family (couple) formation ("gezinsvorming") and family expansion ("gezinsuitbreiding") (NCBS, 1971, 13). However, these two types of patterns could have opposite effects on annual fertility. Marriages were contracted at progressively younger ages. The increasing numbers of young couples resulted in rising numbers of births (first births in particular). On the other hand, changing patterns of family expansion resulted in a reduction of the completed family size of subsequent marriage cohorts, together with decreasing numbers of births. Finally, the NCBS noted that account should be given of patterns of birth intervals, i.e. the spacing of childbearing during marriage.

The arguments given here in favour of an extrapolation of fertility by marriage duration are very general indeed. No empirical evidence is provided. Moreover, they would only be valid under the following necessary (but not sufficient!) condition: there must be a close link between the timing of marriage and the timing of childbearing (in particular the first child) - for instance through a more or less constant first birth interval for subsequent marriage cohorts. Fortunately, this turned out to be the case: first birth intervals for marriages contracted during the 1950s and 1960s showed a stable value of 1.8-1.9 years (NCBS, 1974a, 34; NCBS, 1977, 40).

Frinking and Godefroy, two Dutch demographers who took part in the discussions of the Working Group on Population Forecasting Methodology, provided another empirical argument in favour of marriage duration. They

estimated the completed (marriage) cohort fertility implied by the results of the (age-specific) fertility extrapolations in the 1965 forecast. They found a gradual increase in the mean number of children from a level of 2.8-3.0 children for marriage cohort 1969 to a level of 3.1-3.5 children for marriage cohort 1999 (Frinking and Godefroy, 1967, 18). Because this fertility increase was deemed unrealistic, the authors argued for an approach by marriage duration, instead of age.
How should these considerations be judged, twenty years after the debate took place? The NCBS, in its 1980-based forecasts, abandoned marriage duration and re-introduced age. Three arguments were given (NCBS, 1984, 56-57). First, other forecasters, such as regional forecasters or labour market forecasters often work with age-specific fertility. In these studies, insight in future patterns of age- specific fertility at the national level is often desirable. Second, a methodological argument is that the fertility rates by marriage duration used in the forecasts during the 1970s were calculated as a ratio between births (by marriage duration) and the *initial* size of the marriage cohort ("taux réduit", or "taux par rapport à l'effectif initial", Pressat, 1983, 92). It would be more appropriate to relate those births to the *current* size of the marriage cohort ("taux de première catégorie", Pressat, 1983, 93), because divorce, widowhood and external migration would introduce biases in the former ("reduced") rates. However, data necessary to calculate the latter rates ("of the 1st kind") were not available in the Netherlands in the 1970s and early 1980s. Moreover, the introduction of such rates in the projection model would imply that account should be taken of divorce, widowhood and migration broken down not only by age, but also by marriage duration. The third argument was of an empirical nature: time series analyses revealed that after World War II, trends in fertility rates by marriage duration were no more stable than those in fertility rates by age. In addition, it was expected that in the 1980s and the 1990s, the timing of marriage would be much more loosely connected with the birth of the first child than in the 1950s and the 1960s. This would imply a more stable future trend in the age-specific rates than in the marriage duration-specific rates (NCBS, 1984, 57).

To these three arguments, given by the NCBS, I add the following. The analysis of Frinking and Godefroy must now, twenty years later, be regarded as unconvincing. What they failed to show is, that an extrapolation of fertility by marriage duration, carried out in 1965, would have resulted in much more realistic (declining?) completed cohort fertility levels than the actual age-specific extrapolation. I suppose that this would not have been the case, provided the analysis had been carried out *only* on the basis of information that was available in 1967. Indeed, the mid 1960s were the starting point of a major fertility decline in the Netherlands, which even accelerated in 1970. As in many other Western countries, demographers in the Netherlands did not foresee this dramatic decline, which lasted until 1975. In later chapters I shall demonstrate that the 1970-based forecast, in which fertility was extrapolated by marriage duration, showed larger errors in predicted numbers of births up to 10 years ahead, than the 1965-based forecast did. (For a forecast horizon of 15 years or longer, the reverse is true). Generally speaking, the change in methodology was made during a period in which observed demographic behaviour also changed drastically. This can be regarded as an unfortunate coincidence. As a matter of illustration: the almost perfectly constant length of the first birth interval (1.8 - 1.9 years for marriages contracted until 1965) suddenly disappeared with marriage cohort 1966, when it started to climb steadily to a level of slightly over 3 years for couples who married in 1976 (De Jong, 1987, 14).

4.3 | The introduction of external migration and of variants

The fourth trend in population forecasting methodology in the Netherlands is the gradual incorporation of external migration in the forecasts. In the 5 forecasts with jump-off years 1950, 1951, 1956, 1965 and 1975, the impact of foreign migration on demographic trends is analysed in separate additional calculations. However, these experimental calculations are of an adhoc character; they are presented separately from the natural growth estimates, and each of the publications concerned clearly shows that less emphasis is given to calculations including foreign migration than to those of natural growth only. Therefore, with the exception of the 1975-based forecasts (for reasons to be explained below), my interpretation of those additional computations is that of experimental projections, to which the forecasters did not attach maximum plausibility. Hence, in the remainder of this study, when I refer to the forecasts of 1950, 1951, 1956 and 1965, I have the natural growth estimates in mind.
For the 1975-based forecast the situation is different. Although the main publication consists of two parts - one on natural growth and another on foreign migration, see NCBS (1976) - a few years after it was published the interpretation had clearly shifted. In two articles about the accuracy of the 1975-based forecast the results *including* international migration are seen as *the* forecast results (Kregting, 1979, 1980). From 1980 onwards, international migration was fully incorporated in the forecasts (NCBS, 1982), thereby following advice given as early as 1974 by the Netherlands Scientific Council for Government Policy ("WRR") (WRR, 1974, 34).

Fifthly we may note an increasing use of variants. The 1950 forecast contains two input variants for marital fertility assumptions and two for assumptions on age-specific proportions of married women. Of the theoretical number of 2x2 = 4 output variants, the forecasters only calculated the two extremes. Next, they stated that the most realistic approach would be to use the simple arithmetic average of these two variants, indicated as "Working average". That set of results is considered in this study as the 1950-based forecast. The extreme variants that resulted in the working average are not considered as forecast variants in the sense as defined in chapter 2, because they were not an integral part of the published estimates. The same applies to the 1951-forecast, which consists of a working average based on two extreme output variants, which, in turn, were arrived at by two sets of input assumptions for fertility and two for mortality. Then, for the forecasts of 1956, 1965, 1967 and 1970 no variants were calculated at all. The 1972-based forecast is the first which includes variants as an integral part of the calculations. In the forecasts of 1972 and 1975 the variants were clearly interpreted as "alternative futures" (see section 2.3.1 for a discussion of this notion, and one of the related concept of "uncertainty variants"). For instance, in the low variant of the 1972-forecast ("alternative B", NCBS, 1973, 15) it is assumed that future duration-specific marital fertility rates show a smooth extrapolation of trends observed until 1972. On the other hand, the high variant ("alternative A", NCBS, 1973, 13-14) shows decreasing fertility rates for short marriage durations, and increasing rates for long durations. Although the general trend of fertility was definitely downwards, the latter increase was necessary because the Dutch forecasters wanted to have one variant (A) around replacement level. Hence the high and the low variants were clearly different. For the 1975-forecast the same conclusion can be drawn (NCBS, 1976, 31). But in the 1980-forecast, variants were clearly interpreted as "uncertainty variants" (NCBS, 1982, 17). One set of qualitative future demographic trends was assumed, but in quantifying these trends, more than one variant was introduced for fertility, first marriage and external migration. In fact, the two extreme variants depicted the borders of an area in which a certain trend would take place, perhaps a bit faster, perhaps a little slower, than in the medium variant.

4.4 | Other trends in forecasting methodology

Besides the trends revealed by table 4.1, a number of other trends in forecast methodology in the Netherlands can be observed.
First, it may be noted that, parallel to the increasing interest in assumptions formulated at the general level (see section 4.1), the NCBS gradually paid more attention to the background of these assumptions. In 1974 the first nationwide fertility survey was organized. The sample size was modest (just over 2000 married women were interviewed, see NCBS, 1974b), but in 1977 the second Netherlands Fertility Survey (NFS) involved approximately 7000 married women (Beets and De Graaf, 1980). Sample sizes comparable to that of 1977 were used in the NFS's of 1982, 1985 and 1988. In addition to these surveys organized by the NCBS, the bureau participated in 1975 in the Dutch version of the World Fertility Survey, organized by the NIDI. The results of these surveys were used as one of the inputs to the assumption-making process in the field of fertility: for instance, Van de Giessen (1984, 1985, 1987a, 1987b) studies the usefulness of fertility expectations for forecasting (see also De Beer, 1988). A second illustration,

next to the use of survey data, of a gradually firmer basis for the assumption-making process at various levels is the use of extensive literature reviews undertaken since 1975. This improved the balance between quantitative, analytical demographic aspects of these assumptions on the one hand, and qualitative, sociodemographic aspects on the other. In the forecasts made before 1975, such sociodemographic considerations were virtually absent. In the 1975-based forecast, the NCBS used a number of conclusions formulated by the State Commission for Population Issues regarding future trends in the fields of nuptiality, fertility and mortality (NCBS, 1976, 26-30; State Commission for Population Issues, 1977, 55-60). The assumptions of the forecasts of the 1980s were based on extensive and systematic literature reviews, which resulted in a set of sociodemographic statements relevant for the future demographic developments in the Netherlands (see, for instance, NCBS, 1982, 22ff; Beets and Cruijsen, 1985; Beets and Prins, 1985; Van de Giessen and Van Hoorn, 1985).

A second trend to be noted here applies to the wider context of population forecasting. In the 1970s, two further developments took place. They were largely stimulated by recommendations in 1974 to the government, formulated by the Netherlands Scientific Council for Government Policy (WRR, 1974, 32-36). First, the foundation in 1975 of the Advisory Commission for Population Forecasting, which was to assist the Netherlands Central Bureau of Statistics in making population forecasts. Sitting on this committee are, in addition to the NCBS, representatives of the Central Planning Bureau, the Social and Cultural Planning Bureau (SCP), the National Physical Planning Agency (RPD), the Directorate-General for Housing (DGVH), the Ministry of Education and Science, and the Netherlands Interdisciplinary Demographic Institute (NIDI). Secondly, in 1976 the number of NCBS staff members involved in population forecasting was considerably enlarged (NCBS, 1975, 33).

4.5 | Conclusions

To sum up, we can say that there were two important turning points in the history of post-World War II population forecasting in the Netherlands. First, the year 1970 in which the period age-specific approach to fertility was replaced by a cohort marriage duration-specific method. The second turning point was in 1975, when international migration was introduced into the population forecasting process, and more attention was given to the social scientific basis for forecast assumptions. Moreover, that year also saw significant changes in personnel and organisation.

5. ERRORS IN POST WORLD WAR II NATIONAL DEMOGRAPHIC FORECASTS IN THE NETHERLANDS – A FIRST ANALYSIS

This chapter is the first of a series of three, in which empirical analyses are given of ex-post observed errors in national population forecasts in the Netherlands produced since 1950. The present chapter contains descriptive analyses only, thus providing elements for the error separation model in chapter 6 and for the explanatory model in chapter 7.
I start with a discussion of the data collected for chapters 5-7, the consistency checks that were carried out, the estimation of missing data, and the interpretations which I gave to some issues regarding external migration (section 5.1). Then I present errors in total population size and in annual population growth (section 5.2). Three measures are widely used: the mean percentage error (MPE), the root mean squared percentage error (RMSPE), and (a modification of) Theil's U, of which the definitions were given in section 2.2. Other error measures are defined as they are used. Errors in forecasts of the age structure for males and females are discussed in section 5.3, while components of change are treated in section 5.4. The chapter concludes with a summary of findings (section 5.5), where I try to answer the following questions. Are recent forecasts more accurate than older ones? How fast do errors grow as we look further into the future? And how fast do they diminish as we approach a certain fixed year?

5.1 | Data: checks, estimation of missing information, interpretations and definitions

Table 5.1 presents references to primary sources in which information on NCBS-forecasts can be found. Observed demographic trends are regularly published in the "Maandstatistiek van de bevolking" (Monthly Bulletin of Population Statistics), issued by the NCBS.

Table 5.1. Primary sources for information on NCBS-forecasts

Forecast 1950	1951	1956	1965	1967
NCBS (1951)	NCBS (1954) NCBS (1959)	NCBS (1961) OECD (1961) OECD (1966a) OECD (1966b)	NCBS (1965a) NCBS (1965b) OECD (1966a) OECD (1966b)	NCBS (1967a) NCBS (1967b)
1970	1972	1975	1980	
NCBS (1971)	NCBS (1973)	NCBS (1976)	NCBS (1982) NCBS (1984)	

Total population size by calendar year (1 January) and by forecast is shown in table 5.2, together with observed values. With regard to the 1951-forecast and that of 1956 I was able to trace values for the calendar years, 1956, 1961, 1966, 1971, 1976 and 1981 only. Values for intermediate years were estimated with a procedure to be described below.

As the forecast of 1956 was produced at the end of 1959, observed trends over the years 1956-1958 were known by the time this forecast was released. Hence the results of the 1956-forecast "start" in 1959.

Table 5.3 presents observed numbers of live births by calendar year and corresponding forecasts by calendar year and forecast's jump-off year. Live births for the 1967-forecast for calendar years 1980 and later are neither available in the sources mentioned in table 5.1, nor in archives of the Netherlands Central Bureau of Statistics. However, the archive of the 1967-forecast does contain information on the numbers of 0-year old persons for each calendar year between 1967 and 2000. As the annual numbers of births over the period 1967-1979 show a constant ratio with the corresponding numbers of 0-year old persons at the end of each calendar year ("survival probability"), this ratio was applied to the numbers of 0-year old persons at 1 January of 1981, 1982, ..., 1987 to find the numbers of live births in 1980, 1981, ..., 1986. With respect to the live births of the forecast of 1956 it should be noted that the numbers shown in table 5.3 were never published: because the fertility assumptions of the previous forecast, that of 1951, were close to observed trends, a revision of the 1951-fertility assumptions was considered unnecessary (NCBS, 1959, 18; NCBS, 1961, 657). Thus I copied the numbers of live births of the 1951-forecast to find the corresponding numbers for the 1956-forecast.

Observed numbers of deaths by calendar year and corresponding values from subsequent forecasts are shown in table 5.4. For the 1951-forecast and the 1956-forecast, missing values were estimated. Interpolations carried out by the OECD to find the annual numbers of deaths of the 1956-forecast over the period 1959-1963 resulted in irregular patterns (see OECD, 1966b,

Table 5.2. Total population size, observed and forecasts since 1950, the Netherlands

	Observed	Jump-off year of forecast												
		1950	1951	1956	1965	1967	1970	1972A	1972B	1975A	1975B	1980H	1980L	1980M
1 Jan.						in thousands								
1951	10200.3	10177.7												
1952	10328.3	10308.3	10349											
1953	10435.6	10427.6	10497											
1954	10550.7	10544.9	10644											
1955	10680.0	10659.8	10790											
1956	10821.7	10772.4	10934.7											
1957	10957.0	10882.5	11078											
1958	11095.7	10990.0	11218											
1959	11278.0	11095.1	11357	11285										
1960	11417.3	11197.8	11494	11432										
1961	11556.0	11298.3	11628.4	11576.8										
1962	11721.4	11396.2	11761	11720										
1963	11890.0	11492.1	11891	11862										
1964	12042.0	11586.0	12019	12003										
1965	12212.3	11678.5	12146	12144										
1966	12377.2	11770.4	12272.6	12284.9	12367.8									
1967	12535.3	11860.5	12397	12426	12534.3									
1968	12661.1	11951.4	12524	12569	12705.6	12629.9								
1969	12798.3	12042.6	12652	12715	12882.8	12762.8								
1970	12957.6	12134.3	12781	12864	13066.7	12896.0								
1971	13119.4	12226.0	12911.7	13015.3	13262.3	13034.9	13088.9							
1972	13269.6	12316.9	13042	13168	13464.6	13175.7	13225.8							
1973	13387.6	12409.2	13175	13325	13672.8	13318.1	13366.9	13370.6	13370.6					
1974	13491.0	12501.7	13311	13485	13886.3	13462.3	13510.2	13461.7	13461.7					
1975	13599.1	12593.9	13448	13647	14104.4	13608.0	13655.2	13555.0	13553.9					
1976	13733.6	12685.4	13585.1	13812.1	14331.4	13760.3	13805.0	13670.5	13651.5	13734.9	13734.9			
1977	13814.5	12775.7	13723	13978	14561.3	13913.5	13953.9	13803.9	13749.6	13817.7	13811.9			
1978	13897.9	12864.5	13860	14145	14792.1	14066.7	14102.1	13936.9	13845.4	13893.8	13876.9			
1979	13985.5	12951.2	13998	14313	15024.2	14220.5	14249.8	14066.8	13938.3	13971.5	13938.2			
1980	14091.0	13036.1	14134	14481	15257.9	14375.2	14395.1	14190.2	14028.4	14049.6	13994.2			
1981	14208.6		14270.4	14649.0	15499.4	14537.0	14538.1	14304.3	14116.1	14131.3	14048.2	14211	14211	14211
1982	14285.8				15742.4	14700.1	14679.7	14414.5	14202.0	14216.8	14100.6	14301	14290	14296
1983	14339.6				15985.8	14864.0	14819.5	14521.5	14286.3	14307.8	14154.3	14399	14371	14385
1984	14394.6				16230.2	15028.7	14959.8	14625.5	14368.4	14405.5	14211.0	14499	14448	14473
1985	14453.8				16476.4	15194.8	15097.6	14728.7	14448.8	14508.4	14269.8	14600	14521	14560
1986	14529.4				16730.6	15367.5	15234.0	14830.7	14527.8	14616.0	14331.1	14701	14590	14646
1987	14615.1				16987.5	15541.0	15369.7	14932.2	14606.1	14702.8	14369.8	14801	14658	14730

Table 5.3. Live births, observed and forecasts since 1950, the Netherlands

	Observed	Jump-off year of forecast 1950	1951	1956	1965	1967	1970	1972A	1972B	1975A	1975B	1980H	1980L	1980M
						in thousands								
1950	229.7	222.1												
1951	228.4	212.4	228.6											
1952	231.9	202.8	232.6											
1953	228.0	202.7	233.0											
1954	228.2	202.4	233.1											
1955	229.2	202.0	233.1											
1956	231.5	201.3	232.8											
1957	233.9	200.8	232.0											
1958	236.9	200.2	231.2											
1959	242.5	199.8	230.5	230.5										
1960	239.1	199.6	230.1	230.1										
1961	247.4	198.9	229.0	229.0										
1962	246.2	198.8	228.5	228.5										
1963	249.9	198.7	228.0	228.0										
1964	250.9	199.3	228.3	228.3										
1965	245.2	200.5	229.2	229.2	258.8									
1966	239.6	200.8	229.2	229.2	264.7									
1967	238.7	203.5	232.5	232.5	271.8	234.3								
1968	237.1	205.8	235.4	235.4	280.1	235.1								
1969	247.6	208.3	238.5	238.5	289.1	237.7								
1970	238.9	210.4	241.1	241.1	299.2	241.8	239.0							
1971	227.2	211.6	242.7	242.7	308.3	246.0	248.0							
1972	214.1	215.0	247.5	247.5	316.6	250.0	254.0	214.1	214.1					
1973	195.0	217.3	251.2	251.2	324.2	254.0	259.2	206.2	206.2					
1974	186.0	219.0	254.1	254.1	331.1	257.8	263.9	210.4	209.3					
1975	177.9	220.3	256.6	256.6	337.8	262.1	268.1	235.5	217.2	178.3	178.3			
1976	177.1	221.1	258.4	258.4	343.0	265.5	271.3	256.0	220.3	179.4	173.6			
1977	173.3	221.4	260.1	260.1	346.2	267.7	273.3	257.6	220.2	180.1	168.9			
1978	175.6	221.4	261.3	261.3	349.8	270.5	274.2	256.5	218.8	181.8	165.2			
1979	175.0	221.3	262.3	262.3	353.8	273.9	274.5	252.1	218.3	185.2	162.8			
1980	181.3		263.6	263.6	358.5	277.9	274.6	244.8	218.0	190.0	161.9	181	181	181
1981	178.6				362.4	281.6	274.9	242.8	218.2	195.8	162.4	183	178	181
1982	172.1				365.0	284.5	275.4	241.6	218.5	202.4	164.8	190	183	186
1983	170.2				368.4	287.8	276.0	240.4	218.1	209.3	167.9	195	184	189
1984	174.4				372.5	291.3	276.7	241.4	218.3	215.4	170.7	199	184	192
1985	178.1				377.3	295.0	277.3	242.1	218.6	219.8	172.8	202	186	194
1986	184.5				382.4	298.1	277.9	242.8	219.3	222.4	173.8	204	188	196

Table 5.4. Deaths, observed and forecasts since 1950, the Netherlands

	Observed	Jump-off year of forecast 1950	1951	1956	1965	1967	1970	1972A	1972B	1975A	1975B	1980H	1980L	1980M
					in thousands									
1950	75.9	80.0												
1951	77.6	81.7	83.3											
1952	76.3	83.5	84.6											
1953	80.9	85.3	85.9											
1954	79.6	87.4	87.2											
1955	81.7	89.4	88.5											
1956	84.8	91.2	89.8											
1957	83.0	93.3	91.2											
1958	84.5	95.1	92.6											
1959	86.1	97.1	93.9	84.1										
1960	87.8	99.0	95.4	85.0										
1961	88.3	100.9	96.8	85.7										
1962	94.0	102.9	98.3	86.4										
1963	95.7	104.8	99.7	87.0										
1964	93.4	106.7	101.2	87.6										
1965	98.0	108.6	102.8	88.1	95.9									
1966	100.5	110.6	104.3	88.5	98.2									
1967	99.8	112.5	105.9	88.9	100.5	100.1								
1968	105.0	114.6	107.5	89.3	102.9	102.3								
1969	107.6	116.7	109.1	89.6	105.2	104.5								
1970	109.6	118.7	110.8	90.0	103.6	102.9	109.6							
1971	110.2	120.6	112.4	90.3	106.0	105.3	111.1							
1972	113.6	122.7	114.1	90.7	108.4	107.5	112.9	113.6	113.6					
1973	110.7	124.8	115.7	91.0	110.7	109.8	115.9	115.0	115.0					
1974	109.3	126.7	117.4	91.4	113.0	112.1	118.9	117.1	117.1					
1975	113.7	128.8	119.1	91.9	110.7	109.9	118.3	120.0	119.8	115.3	115.3			
1976	114.5	130.7	120.8	92.4	113.1	112.3	122.4	122.4	122.0	117.4	117.3			
1977	110.1	132.6	122.4	93.0	115.4	114.5	125.1	124.6	124.1	119.5	119.4			
1978	114.4	134.7	124.1	93.7	117.7	116.8	126.5	126.7	126.2	121.7	121.5			
1979	112.6	136.4	125.7	94.4	120.1	119.1	129.2	128.6	128.2	123.8	123.5			
1980	114.3		127.4	95.3	117.0	116.1	131.6	130.7	130.3	124.1	123.8	114.0	114.0	114.0
1981	115.5				119.4	118.5	133.3	132.6	132.3	126.3	125.9	116.0	116.0	116.0
1982	117.3				121.6	120.7	135.6	134.6	134.2	128.5	128.1	118.0	117.0	117.0
1983	117.8				124.0	123.0	135.7	136.4	136.0	130.6	130.2	120.0	120.0	120.0
1984	119.8				126.3	125.2	138.9	138.2	137.9	132.8	132.2	122.0	121.0	121.0
1985	122.7				123.1	122.3	140.9	139.8	139.5	133.7	133.1	123.0	123.0	123.0
1986	125.3				125.5	124.7	142.2	141.6	141.2	135.7	135.2	125.0	125.0	125.0

182), and therefore were not used here. A description is given of the estimation procedure applied for the annual numbers of deaths in the 1951-forecast. Estimations for the 1956-forecast were carried out following a similar approach.

Forecasts for annual numbers of live births, as well as those for total population size at 1 January of the years 1961, 1966, ..., 1981 were available from sources listed in table 5.1. Because international migration was not considered in the forecast of 1951 (see table 4.1), total numbers of deaths over the period 1956-1960, 1961-1965, ..., 1976-1980 could easily be derived from the available information. Hence the estimation procedure to be used would simply have to break down those quinquennial numbers into annual numbers. A fourth degree polynomial function (of time) was assumed to describe the annual pattern, and the five parameters of this function were calculated with the aid of the five data points for the periods 1956-60, 1961-65, ..., 1976-80. With known parameters the polynomial function was employed to give estimates of annual numbers of deaths over the period 1956-1980 (interpolations), as well as 1951-1955 (extrapolations). Having estimated the annual deaths of the 1951-forecast and the 1956-forecast, estimates of missing total population sizes for these forecasts were derived with the help of known annual numbers of births and of known total population sizes for the years 1951, 1956, 1961, ..., 1981.

Table 5.5 contains observed numbers of net-immigrations for the years 1950-1986 and corresponding numbers for the forecasts of 1975 and 1980. As international migration was not considered in NCBS-forecasts produced before 1975, my interpretation is that the forecasters implicitly assumed a net-immigration of zero as the most plausible trend in the pre-1975 forecasts. This interpretation is consistent with the definition of a population forecast employed throughout this book.
Net-migration figures for the 1975-forecast are those labeled as "alternative 1" in the publication (NCBS, 1976, 111). The variant labeled as "alternative 2", showing a net-deficit, was not considered very plausible (Kregting, 1979, 1980). Hence it is not included in this study.

In order to ascertain the internal consistency of the data set, the following checks were made. For the observed population the following relation must hold: population size at 1 January of year t+1 equals population size at 1 January of year t plus the number of live births during t minus the number of deaths during t plus the net-number of immigrants during t plus the net-number of administrative corrections. The last term is included in this relation, because observed NCBS-numbers apply to the "de jure" population of the Netherlands. Time lags in registration and misreporting of births, deaths and migration make these administrative corrections necessary. They are generally small in number, most often caused by emigration without reporting ("destination unknown"). Exceptions are the years 1958 and 1963. In

Table 5.5. Net-migration, observed and forecasts since 1950, the Netherlands

	Observed	Jump-off year of forecast 1950	1951	1956	1965	1967	1970	1972A,B	1975A,B	1980H	1980L	1980M
				in thousands								
1950	19.9	0.0										
1951	-22.7	0.0	0.0									
1952	-47.7	0.0	0.0									
1953	-31.6	0.0	0.0									
1954	-19.0	0.0	0.0									
1955	-5.2	0.0	0.0									
1956	-11.0	0.0	0.0	0.0								
1957	-12.5	0.0	0.0	0.0								
1958	12.1	0.0	0.0	0.0								
1959	-16.9	0.0	0.0	0.0								
1960	-12.8	0.0	0.0	0.0								
1961	6.4	0.0	0.0	0.0								
1962	17.4	0.0	0.0	0.0								
1963	8.4	0.0	0.0	0.0								
1964	13.7	0.0	0.0	0.0								
1965	18.8	0.0	0.0	0.0	0.0							
1966	20.0	0.0	0.0	0.0	0.0							
1967	-11.5	0.0	0.0	0.0	0.0	0.0						
1968	6.0	0.0	0.0	0.0	0.0	0.0						
1969	20.2	0.0	0.0	0.0	0.0	0.0						
1970	33.5	0.0	0.0	0.0	0.0	0.0	0.0					
1971	33.0	0.0	0.0	0.0	0.0	0.0	0.0					
1972	19.1	0.0	0.0	0.0	0.0	0.0	0.0	0.0				
1973	21.1	0.0	0.0	0.0	0.0	0.0	0.0	0.0				
1974	33.1	0.0	0.0	0.0	0.0	0.0	0.0	0.0				
1975	72.1	0.0	0.0	0.0	0.0	0.0	0.0	0.0				
1976	21.4	0.0	0.0	0.0	0.0	0.0	0.0	0.0	20.6			
1977	22.8	0.0	0.0	0.0	0.0	0.0	0.0	0.0	15.4			
1978	28.1	0.0	0.0	0.0	0.0	0.0	0.0	0.0	17.4			
1979	44.8	0.0	0.0	0.0	0.0	0.0	0.0	0.0	16.6			
1980	53.0		0.0	0.0	0.0	0.0	0.0	0.0	15.8			
1981	17.0			0.0	0.0	0.0	0.0	0.0	15.9	23	17	20
1982	3.2			0.0	0.0	0.0	0.0	0.0	17.0	25	16	20
1983	6.0			0.0	0.0	0.0	0.0	0.0	18.9	25	13	19
1984	8.1			0.0	0.0	0.0	0.0	0.0	20.2	24	10	17
1985	24.1			0.0	0.0	0.0	0.0	0.0	21.4	22	7	15
1986	32.7			0.0	0.0	0.0	0.0	0.0	0.0	21	5	13

1958, about 18 thousand Amboynese, living in temporary dwellings, and kept outside of the regular registration system since the beginning of their stay in the Netherlands, were included in the population registers and hence in the population figures. In 1963, a correction of the frontier between the Netherlands and the FRG took place, as a consequence of which about 10 thousand inhabitants of the villages of Elten and Tudderen became FRG-citizens. For the forecast demographic numbers the above-mentioned relation must also hold, with the exception of the component of administrative corrections, as this term is not included in the NCBS-forecasts under consideration. Indeed, except for rounding-errors, the two relations hold exactly. Tables 5.2-5.5 present the greater part of the input data used in this study. Additional data (such as age-specific fertility rates, age-specific proportions of married women, duration-specific fertility rates, numbers of females marrying, life expectancies) are employed occasionally. They will be discussed in the sections where they are first introduced.

Finally, some words on definitions. All stock variables in this study are measured at 1 January of the relevant calendar year. Flow variables apply to the period 1 January-31 December of the calendar year in question. Flow variables and parameters that are specified by age or by age-group are defined over a period-cohort observation criterion (i.e. they follow cohorts over the relevant period), unless specified otherwise.

5.2 | Errors in total population size and in population growth

Table 5.6 presents an overview of the accuracy of post-war forecasts of the total population for the Netherlands. The forecasts of the early 1950s, as measured by the mean percentage error (MPE) of annual forecast results show, on the whole, an underestimation. The same can be observed for the low variant of the 1972-forecast, and for the 1975-forecast (both variants). To enable a comparison between forecasts, MPE's are shown over a forecast period of 1-5 years, 1-10 years and 1-15 years. In general, short-term (1-5 years) and medium-term (1-10 years) forecasts have slightly overestimated total population size, with 0.04 per cent and 0.05 per cent respectively. The overestimation is, however, substantially larger for longer periods: 0.60 per cent for 1-15 years and (not shown in table 5.6) 0.64 per cent for 1-20 years. The 1956-forecast shows relatively small errors, as does the high variant of the 1975-forecast. By far the largest deviations are found for the 1965-forecast.
Since overestimations and underestimations compensate each other, the lower panel of table 5.6 contains results on the accuracy of total population size forecasts measured by the root mean squared percentage error (RMPSE). A comparison between corresponding entries of the lower panel and the upper panel shows that such compensation effects did occur indeed. The drop in accuracy with increasing length of the forecast period is much

Table 5.6. Accuracy of total population size forecasts since 1950, the Netherlands

	Jump-off year of forecast							
	1950	1951	1956	1965	1967	1970	1972H	1972L
Forecast period					MPE (%)			
1-5 yrs.	-0.15	0.75	-0.05	0.35	-0.47	-0.03	-0.24	-0.35
1-10 yrs.	-0.64	0.80	-0.38	1.31	-0.21	0.69	0.19	-0.41
1-15 yrs.	-1.53	0.45	-0.44	3.00	0.53	1.58		

	1975H	1975L	1980H	1980M	1980L	All forecasts
					MPE (%)	
1-5 yrs.	0.08	-0.24	0.45	0.34	0.22	0.04
1-10 yrs.	0.12	-0.75				0.05
1-15 yrs.						0.60

	1950	1951	1956	1965	1967	1970	1972H	1972L
Forecast period					RMSPE (%)			
1-5 yrs.	0.16	0.82	0.20	0.50	0.51	0.27	0.28	0.39
1-10 yrs.	0.90	0.84	0.53	1.78	0.46	1.10	0.51	0.44
1-15 yrs.	2.09	0.74	0.56	4.04	1.27	2.18		

	1975H	1975L	1980H	1980M	1980L	All forecasts
					RMSPE (%)	
1-5 yrs.	0.14	0.35	0.59	0.43	0.28	0.42
1-10 yrs.	0.29	0.92				0.88
1-15 yrs.						2.16

more regular in the case of RMSPE than in the case of MPE, as shown in the column labeled "All forecasts". On the whole, the 1965-forecast shows a bad performance here too, but that of 1951 was worse on the short run (1-5 years).

As a first step towards an analysis of accuracy of demographic components (birth, death, migration, see section 5.4) I present in table 5.7 some accuracy measures for forecasts of the growth of the total population in the Netherlands since 1950. Relative growth in year t was calculated as $(P_{t+1}-P_t)/P_t$, with P_t the population size at time t. The first line of table 5.7 shows the correlation between observed growth and forecasted growth. In six out of thirteen forecasts, the coefficient is negative. Because population growth in the Netherlands was positive during the entire period 1950-1986 (both actual and pro-

Table 5.7. Accuracy of total population growth forecasts since 1950, the Netherlands

	Jump-off year of forecast													
	1950	1951	1956	1965	1967	1970	1972H	1972L	1975H	1975L	1980H	1980M	1980L	All forecasts
	Correlation between observed and forecasted growth													
	0.46	0.43	0.12	-0.58	-0.70	0.65	0.22	0.65	-0.68	0.03	-0.50	-0.90	-0.57	
	Error in average annual growth rate (percentage points)													
ME	-0.22	0.03	-0.06	0.40	0.12	0.14	0.04	-0.06	-0.01	-0.11	0.14	0.10	0.06	0.03
RMSE	0.25	0.13	0.10	0.46	0.18	0.21	0.11	0.07	0.05	0.13	0.17	0.12	0.07	0.20
SDE	0.11	0.11	0.10	0.22	0.13	0.16	0.10	0.05	0.05	0.06	0.07	0.05	0.04	0.12
MAE	0.22	0.09	0.10	0.40	0.15	0.19	0.10	0.06	0.04	0.11	0.14	0.10	0.06	0.15

jected growth), this indicates a weak growth in the forecast in periods when actual growth was strong, and vice versa. For the remaining seven cases with positive correlation coefficients, the relationship between observed and forecasted growth is often quite weak, see, for instance, the forecasts of 1956, 1972 (high variant) and 1975 (low variant).

On the second and the third line I give statistics of errors in average annual growth rates according to the approach employed by Keyfitz (1981) and Stoto (1983). These two authors compared observed and forecasted growth as follows. If P_t represents total population size at time t, then the average annual growth rate r over the period (0, T) is calculated as

$$r = \{(P_T \ / \ P_0)^{1/T}\} - 1.$$

An analysis of the difference Δr between forecasted r and observed r allows one to draw conclusions about the accuracy of growth estimates. Lines 2-5 show the mean of the error Δr (ME), its root mean squared value (RMSE), its standard deviation (SDE) and its mean absolute value (MAE).
The slightly positive value of ME for all forecasts together implies that, on the whole, forecasts over the period 1950-1986 were a little too high, overestimating real population growth. The major factors determining this were the forecasts of 1965, 1967 and 1970.
In general, the results of this analysis tell us that the more recent forecasts do not show smaller errors, and that an important factor determining the error is the date the forecast was made. The latter agrees with the findings of Keyfitz and Stoto, but these authors *did* find an improvement over time of forecast accuracy. Moreover, the overall value of the RMSE of 0.20 percentage points reveals that in the Netherlands the accuracy of forecasts of population growth was somewhat higher than in comparable countries, for which Keyfitz found an RMSE of 0.29 percentage points. Stoto computed a standard deviation of 0.50 percentage points for US-forecasts made after 1945. He gives two assumptions for the standard deviation in the growthrate of future US-forecasts: an optimistic one of 0.3 and a pessimistic one of 0.5 (Stoto, 1983, 17). The value of the standard deviation for the Netherlands of only 0.12 percentage points shows that Dutch forecasters were more successful over the period after 1945 than their colleagues in the USA. A similar conclusion can be drawn from a comparison of RMSE's for the United States (Long, 1987, table 1) and those for the Netherlands in table 5.7. Long shows that the RMSE in USBC growth rate forecasts varies between 0.2 and 1.1 percentage points for the forecasts produced between 1945 and 1970. Relatively large errors in population growth are exhibited by the USBC-forecasts of 1945 (1.1 %), 1957 (0.6 %) and 1966 (0.5 %).

5.3 | Errors in age structure

Most users of population forecasts are not (only) interested in total population size, but in specific age-groups (as well). Hence it is useful to look at errors in forecasts of the age structure of the population. Indeed, the accuracy of forecasts of the population differ by age group, as we shall see in this section.

Forecast results by sex and five-year age groups are available for all forecasts made by the NCBS since 1950. They were brought together by Schreurs (1989, Appendix B) in an extensive set of tables. Results from the forecasts of 1965 and 1967, for each of the calendar years 1965-1987 were obtained from NCBS-archives. Other data were published in the relevant forecast publications. Unfortunately, information on the age structures of the 1951-forecast and the 1956-forecast is only available for the years 1956, 1961, 1966, ..., 1981. For these two forecasts the highest open-ended age group is 75+, whereas I have information for all other forecasts up to age 85+ (and sometimes even for higher ages).

Figure 5.1 shows the MPE for the age structure of the population. Contrary to table 5.6, the errors per age group and per forecast have been calculated for a forecast period of exactly 5 years, 10 years, and 15 years. I took the mean of the different forecasts: all 13 forecasts for a 5-year period given in table 5.6, the 10 forecasts prior to and including that of 1975 for 10 years ahead and the 8 forecasts before 1975 for a 15-year period. Figure 5.1 clearly shows the magnitude of overestimation for the youngest age group: up to 25% for 0- to 4-year-olds for the 15-year forecasts. We will see later that this can be attributed mainly to an overestimation of fertility in the Dutch forecasts made between 1965 and 1972. An underestimation of the elderly population results from a too pessimistic evaluation regarding mortality. As the length of the forecast period rises, so do the forecast errors. However, for ages between 10 and 74 years, the maximum over- or underestimation is 2%. At more advanced ages, forecast errors become larger, especially if the forecast period exceeds 10 years. Underestimations of similar size for the elderly population have been found for Denmark, Norway and the United Kingdom (Leeson, 1981, 77; Brunborg, 1984, 14; Field, 1988). This goes to show that for ages over 75, it is relatively difficult to make an accurate forecast.

Differences between males and females in MPE-patterns may be attributed to sex-differences in external migration and in mortality. For instance, at age 30-34 male MPE's are 0.5 to 1.5 percentage points larger in absolute value than female MPE's. This is caused by immigration of labourers in the 1970s which was higher than expected. But at ages 45-55 the MPE's for females are about 1 percentage point higher than those for males: this is due to unforeseen family reunion. Finally, for ages beyond 80, female errors are about 2 (5 years ahead) to 10 (15 years ahead) percentage points higher than male errors. In section 5.4 we shall see that mortality forecasts made in the 1950s, 1960s and early 1970s were structurally worse for females than for males.

The general shape of figure 5.1 is characteristic for a country with rapidly decreasing fertility and a more or less steady improvement of mortality.

Keilman and Kucera (forthcoming) found for Czechoslovakia a pattern with relatively high errors at low and at high ages, and errors which grow as the forecast horizon becomes further removed from the jump-off year. But the Czechoslovakian errors at low ages are much smaller than those for the Netherlands. The reason is that actual births observed in Czechoslovakia between 1950 and 1985 show fluctuations around a more or less constant level. Moreover, relatively large errors are observed in the Czechoslovakian forecasts for males above age 60: they start at 2-3 per cent for the age group 60-64, and increase regularly to 12 per cent for a forecast horizon of 10 years and 17 per cent for 15 years ahead. Indeed, male mortality levels were

Figure 5.1. Mean percentage error of forecasts of the age structure since 1950, the Netherlands

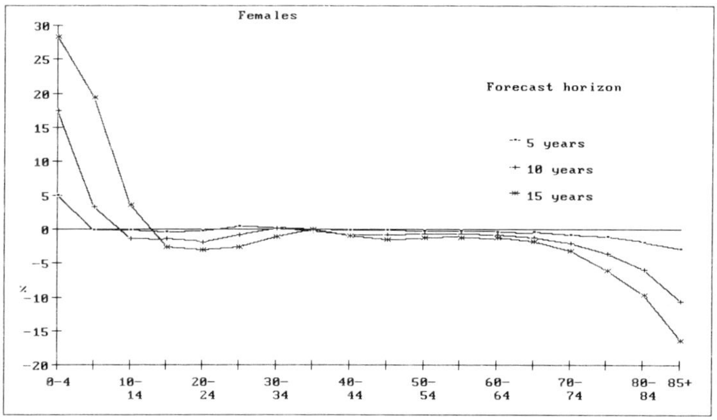

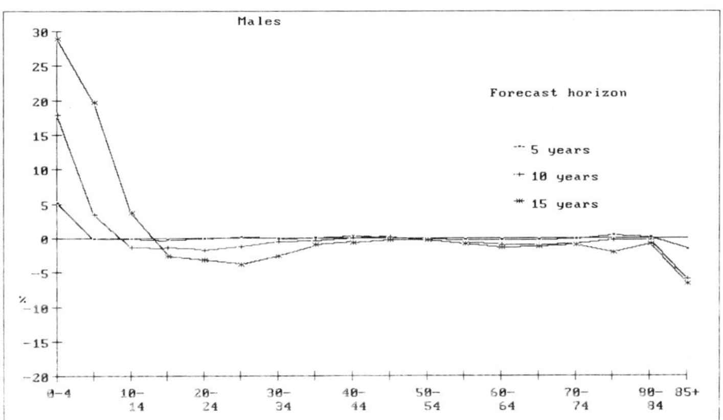

underestimated in Czechoslovakia, in particular in the 1960s (Keilman and Kucera, forthcoming).
Field (1988) investigated forecasts with jump-off years 1971, 1976 and 1981 for the United Kingdom and compared the age structure with the actually observed figures in 1976, 1981 and 1986. The shape of the UK-error curve is the same as that for the Netherlands: at ages 0-4 Field finds average overestimations of 4 per cent (5 years ahead), 21 per cent (10 years) and 32 per cent (15 years). At ages 85 and over the figures show an average underestimation of 4 per cent (5 years ahead), 8 per cent (10 years) and 17 per cent (15 years).
In Finland, fertility increased in the beginning of the 1980s, and life chances improved in that period more slowly than expected. Therefore, Hämäläinen (1987, 8) found for a large set of municipal forecasts made in 1981 a pattern with large mean negative errors for low age-groups, with large mean positive errors for high age-groups (80+), and with only minor errors for intermediate ages.
Finally, Sabatello (1988, 285) studied the accuracy of the forecasts of the population aged 75 and over in Israel, as they were produced in 1965, 1970 and 1975. In 1985 the observed proportion of this age group in the total population was 3.5 per cent, and this was higher than was foreseen for that year in the forecasts of 1965 (2.9 per cent), 1970 (3.1 per cent) and 1975 (3.2 per cent). Much of the discrepancies stem from an unexpected immigration of the elderly in the late 1970s.
Figure 5.2 shows the RMSPE in the age structure. A comparison with figure 5.1 reveals that, in particular, for generations not yet born at the time the forecast is made errors are extremely high: up to 45 per cent for the youngest (0-4 years) after 15 years! The corresponding MPE in figure 5.1 is "only" 27 per cent, which reveals strong compensation effects of over- and underestimations.

Figure 5.2. Root mean squared percentage error of forecasts of the age structure since 1950, the Netherlands

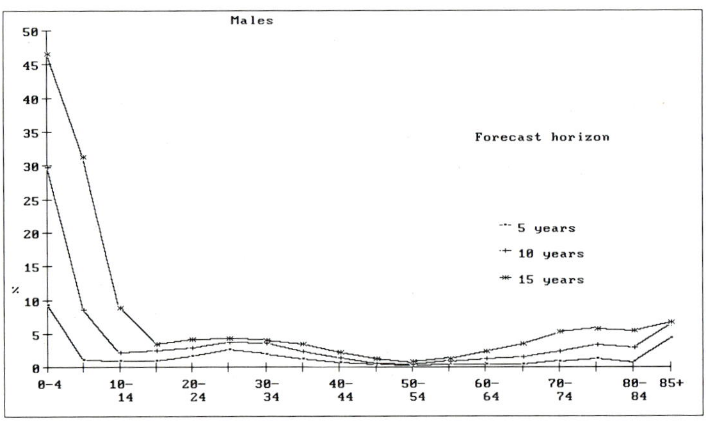

5.4 | Errors in components of change

Most forecast *users* interested in the accuracy of past population forecasts will be satisfied with error statistics for the total population and the age structure. Some of them will also be interested in error patterns for the components of change: absolute numbers of births, of deaths and of migrants. However, if one is interested in the *methodology of population forecasting*, one will want to know more, and descend from the level of forecast results to the level of assumptions. It is not possible to make an error analysis at the most detailed level, namely the level of age-specific assumption parameters (level 1, see section 4.1), since this information is usually no longer available for most forecasts made in the past. Moreover, one may expect to find the major patterns in forecast errors at the next highest level (level 2), that is the level of summary variables, such as the total fertility rate, the life expectancy, etc. These and other summary variables bring together in one number the separate age-specific rates.

In this section I present comprehensive results on error patterns in absolute numbers of births, deaths and migrants, in crude birth rates and crude death rates, and some findings on summary variables for fertility, mortality and international migration.

5.4.1. Fertility

Table 5.8 shows that in 1965, the short-term (up to 5 years) fertility forecast was better than those in 1970 and 1972. That of 1967 was very accurate, because the birth decline which set in in 1964 in the Netherlands was interrupted in 1969 by a substantial increase in fertility (see figure 5.3). In 1970, fertility started to drop drastically. The 1967 fertility forecast was, on average, fairly accurate for the first 5 years, since the first three years had been underestimated, and the last 2 years overestimated. However, Theil's U shows that a naïve short-term forecast keeping absolute numbers or crude rates constant, would have been better, both in 1967 and in 1970. Since the observed birth decline between 1964 and 1970 was much less dramatic than that after 1970, the 1965 short-term fertility forecast contained smaller errors than the 1970- and 1972-forecasts. For the ten-year forecast period, the 1965-fertility forecast was again better than the 1970-forecast. Only when looking 15 years ahead, does the 1970-fertility forecast become more accurate. As shown in the above, the fertility forecast of 1967, measured by the MPE and the RMSPE is still significantly better than the 1970-forecast.

How do these findings for the Netherlands compare to errors in other countries? In general, the accuracy of Dutch fertility forecasts is not less than that in the United States. The RMSPE for 5-year fertility forecasts in the Netherlands which fluctuates between 1% and 26%, is of the same magnitude as the corresponding RMSPE that Ahlburg (1982, 368) found for post-war fertility forecasts in the United States: a minimum RMSPE of 2% for the 1967 US-forecast, and a maximum of 21% for that of 1970. For the

Figure 5.3. Live births, forecasts and observations, the Netherlands

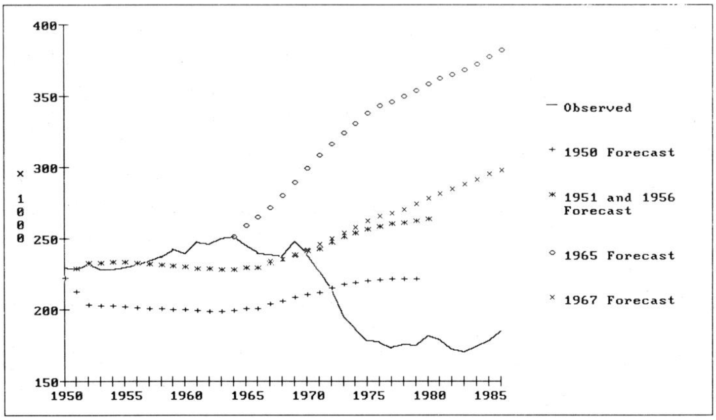

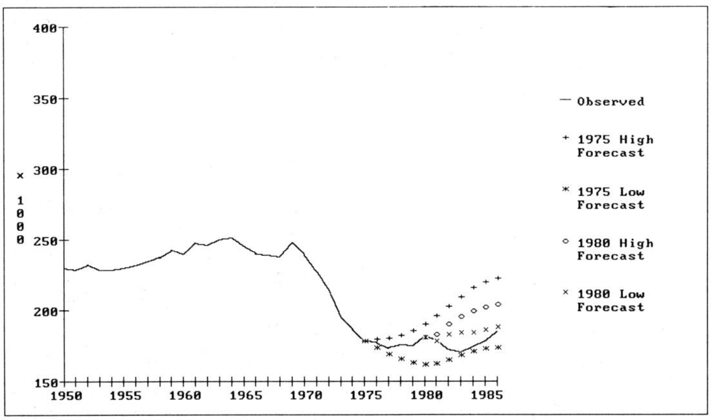

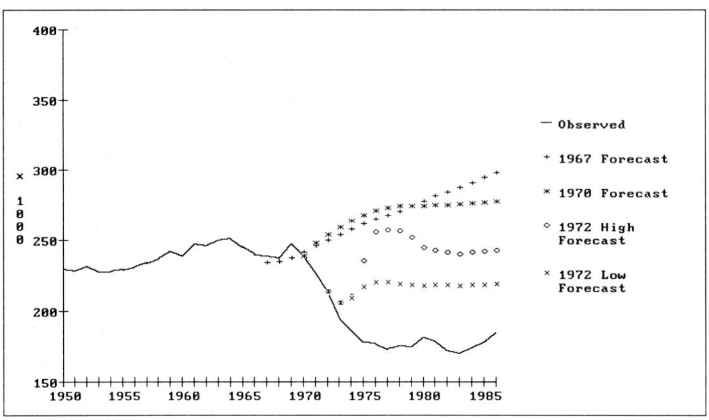

Table 5.8. Accuracy of fertility forecasts, by length of forecast period, the Netherlands

	Jump-off year of forecast													
	1950	1951	1956	1965	1967	1970	1972H	1972L	1975H	1975L	1980H	1980M	1980L	Total
					5 year period									
Number of live births														
MPE (%)	-9.0	1.3	-6.4	13.0	0.6	20.5	19.2	13.0	3.0	-3.4	8.3	6.1	3.9	5.4
RMSPE (%)	9.7	1.6	6.7	13.7	4.3	25.6	25.5	16.0	3.6	4.3	10.2	7.6	5.2	12.8
U	3.1	2.3	1.8	3.3	1.5	1.2	1.1	0.7	0.6	0.7	4.3	3.2	2.2	1.2
Crude birth rate														
MPE (%)	-9.0	0.7	-6.4	12.7	0.9	20.7	19.5	13.3	3.0	-3.3	8.0	5.9	3.8	5.4
RMSPE (%)	9.6	1.0	6.7	13.5	4.5	25.6	25.8	16.4	3.6	4.2	9.9	7.3	5.1	12.8
U	1.2	0.2	4.9	1.6	0.8	1.0	1.0	0.6	0.4	0.5	2.8	2.1	1.4	1.0
					10 year period									
Number of live births														
MPE (%)	-11.7	-0.5	-5.5	31.8	18.6	37.7	30.6	18.4	9.3	-4.5				12.4
RMSPE (%)	12.4	2.4	6.1	39.5	27.4	42.9	34.9	20.3	12.5	5.6				24.8
U	4.8	1.0	1.8	2.5	1.4	1.3	1.4	0.8	2.0	0.9				1.4
Crude birth rate														
MPE (%)	-11.4	-1.2	-5.2	30.4	18.9	37.0	30.4	18.8	9.5	-3.9				12.3
RMSPE (%)	11.9	2.7	5.9	37.4	27.5	41.9	34.7	20.8	12.7	5.1				24.3
U	1.2	0.4	0.9	1.5	1.0	1.0	1.1	0.6	1.2	0.5				1.1
					15 year period									
Number of live births														
MPE (%)	-14.3	-2.9	-0.5	53.5	30.8	44.2	32.9	20.4						20.5
RMSPE (%)	15.1	4.9	10.0	64.7	39.0	48.2	35.9	21.8						35.6
U	4.4	1.1	1.5	2.3	1.5	1.3	1.3	0.8						1.6
Crude birth rate														
MPE (%)	-13.2	-3.4	-0.0	49.2	30.2	42.3	32.1	20.8						19.7
RMSPE (%)	13.8	5.1	10.2	58.7	37.9	45.8	35.0	22.3						33.6
U	1.2	0.6	0.6	1.4	1.0	1.0	1.0	0.6						1.0

forecasts over a period of 10 years and 15 years, the RMSPE's for fertility in the Netherlands are also comparable to those in the United States. And Theil's U does not show much difference among the two countries, although Ahlburgs analysis indicates that the range of U in the US is somewhat narrower than that in the Netherlands: between 0.65 and 4.08 for the US, between 0.20 and 4.87 for the Netherlands. Finally, it may be noted that the 5-year mean percentage forecast errors for Dutch live births have the same order of magnitude as those for the Norwegian forecasts of live births made since 1969 (Brunborg, 1984, 91).

So far I have discussed errors in absolute numbers of births and in crude birth rates. However, forecast assumptions are not formulated in terms of these variables, but rather in the form of summary indicators, or of detailed parameters. I shall now analyse errors in summary indicators for fertility, which enables me to investigate the performance of this demographic component without the disturbing effect of over- or underestimations in child-bearing ages due to inaccurate forecasts for mortality and migration. Information on detailed parameters in historical forecasts is difficult to obtain. Therefore the analysis must be restricted to the summary level.
I limited the number of forecasts to be investigated at this level to 3: those of 1950, 1965 and 1970. Availability of data was one important argument for the selection. Another was the intention to have some methodological variation within the set of selected forecasts. Van Poppel (1984) collected much material for the 1965-forecast. Hence it was natural to add the 1970-forecast, given the methodological change in fertility between these two forecasts. The 1950-forecast was added in order to cover not too short a period.
I calculated annual mean numbers of marital live births per woman for the forecasts of 1950, 1965 and 1970. For the forecasts of 1950 and 1965 this was straightforward. Annual marital fertility rates in 5-year age groups (under 20, 20-24, 25-29, ..., 45-49) and corresponding proportions of married women are available for the 1950-forecast (1950-1979, see NCBS, 1951, p. 29, p. 35) and for that of 1965 (collected by Schreurs, 1989, p. 110, p. 111 from unpublished NCBS-files). The sum over all age groups of the product between marital fertility rate and proportion married women (multiplied by five) resulted in a period (synthetic) mean number of marital live born children per women for each calendar year. With respect to the forecast of 1970, published cohort-assumptions on marital fertility rates by year of marriage (1952-1999), marriage duration (0, 1, 2, ..., 19, 20+) and age at marriage (under 20, 20-24, 25-29, 30-49) (see NCBS, 1971, 76-79, supplemented with information from NCBS-files), were translated into period-assumptions on marital fertility rates by calendar year (1970-1986), year of marriage (1952-1986) and age at marriage (under 20, 20-24, 25-29, 30-49). For each age at marriage, the sum of all fertility rates in a certain calendar year resulted in a period (synthetic) mean number of children for that year. By taking a weighted average (with weights equal to forecasted num-

Table 5.9. Mean number of marital live births per woman, forecasts and observed (period perspective), the Netherlands

	Observed		Jump-off year of forecast		
	age-specific	marriage duration-specific	1950	1965	1970
1949	3.16				
1950	3.02		3.02		
1951	2.99		2.86		
1952	3.03		2.71		
1953	2.98		2.69		
1954	2.98		2.67		
1955	2.98		2.65		
1956	3.00		2.62		
1957	3.02		2.61		
1958	3.04		2.59		
1959	3.11		2.57		
1960	3.06	2.75	2.55		
1961	3.14	2.83	2.54		
1962	3.10	2.79	2.52		
1963	3.11	2.81	2.50		
1964	3.07	2.77	2.49		
1965	2.94	2.64	2.47	3.14	
1966	2.83	2.51	2.46	3.16	
1967	2.72	2.42	2.44	3.15	
1968	2.63	2.33	2.43	3.15	
1969	2.67	2.37	2.41	3.17	
1970	2.51	2.23	2.40	3.19	2.31
1971	2.33	2.06	2.38	3.23	2.29
1972	2.13	1.94	2.37	3.22	2.29
1973	1.89	1.73	2.36	3.22	2.29
1974	1.75	1.63	2.34	3.22	2.29
1975	1.64	1.56	2.33	3.23	2.30
1976	1.61	1.57	2.31	3.24	2.29
1977	1.56	1.55	2.30	3.24	2.29
1978	1.57	1.59	2.29	3.24	2.29
1979	1.54	1.62	2.28	3.24	2.28
1980	1.56	1.71		3.25	2.28
1981	1.51	1.70		3.25	2.26
1982	1.43	1.65		3.25	2.26
1983	1.40	1.65		3.26	2.25
1984	1.41	1.68		3.26	2.25
1985	1.42	1.72		3.26	2.24
1986	1.44	1.77		3.27	2.24

bers of marrying females by age at marriage) of these means the period mean number of marital live births per woman for each of the years 1970-1986, irrespective of age at marriage was found. A similar procedure resulted in corresponding observed mean numbers of children for the period 1970-1986. Table 5.9 contains summary fertility indicators for the forecasts of 1950, 1965 and 1970, together with corresponding observed values.

Error statistics regarding the mean number of marital live births per woman for the three forecasts are shown in table 5.10. There is a strong parallel with the accuracy results regarding absolute numbers of live births and crude birth rates (compare table 5.8). In the short run (1-5 years), the forecast of 1965 shows smaller errors than that of 1970, but the situation reverses for longer forecast periods (10 years or more). A difference with table 5.8 is that at the level of summary indicators, the 1970-forecast is 20-40% better than a naïve forecast (U is between 0.6 and 0.8). At the level of absolute numbers and crude rates it is up to 50% worse than a naïve forecast (with a U between 1.0 and 1.5). The reason is that the fertility decrease between 1969 (the level of fertility in this year is taken as a naïve forecast) and 1975 (when the downward trend stabilized) is much stronger in terms of the mean number of marital live births than in terms of absolute numbers or crude rates.

Table 5.10. Accuracy of mean number of marital live births per woman, the Netherlands

	Jump-off year of forecast											
	1950				1965				1970			
	Length of forecast period											
	1-5	1-10	1-15 (years)	total	1-5	1-10	1-15 (years)	total	1-5	1-10	1-15 (years)	total
MPE (%)	-7.1	-10.5	-13.2	1.3	14.6	34.4	57.7	79.0	19.7	32.6	33.3	32.7
RMSPE (%)	8.3	11.5	14.3	23.0	15.3	42.5	69.6	90.9	24.8	36.6	36.0	35.2
U	1.5	2.3	3.4	0.6	1.2	1.2	1.1	1.1	0.7	0.8	0.8	0.6

As was the case with total births, there is a strong similarity with the pattern in the United States when one evaluates forecasts of fertility summary indicators. Alho (1985, figure 1) presents actual Total Fertility Rates for the period 1920-1980, and projected TFR's for the forecasts of 1945, 1957, and 1963. The forecast of 1945 comes out far too low. The situation is similar to that for the 1950 forecast for the Netherlands, although the underestimation in the USA is much stronger than that in the Netherlands. The reason is that there was a rapid increase in fertility in the USA between 1945 and 1960 (from approximately from 2.5 to 3.5 in that period), whereas table 5.9 indicates that fertility in the Netherlands was fairly stable between 1950 and 1965. On the other hand, the fertility forecasts for the USA made in 1957 and 1963 resulted in strong overestimations, because the actual TFR declined rapidly between 1956 and 1973. This is analogous to the situation in the Netherlands for the forecasts of 1965 and 1970, as can be seen from table 5.9.

The summary indicators are composed of a marital fertility component and a nuptiality component. An obvious question is: which of the two components contributed most to forecast errors? One way to answer this question is with direct standardization: a comparison between a series of *observed* mean numbers of children and a series obtained by multiplying *observed* fertility age-specific marital rates with *assumed* proportions married women, would indicate to what extent errors in nuptiality assumptions have caused errors in mean numbers of children in the forecast of 1950 and that of 1965. When observed figures are taken as the standard, one standardizes for fertility in this case. The combination of *assumed* fertility rates and *observed* proportions married women tells us to what extent marital fertility assumptions are responsible for errors in mean numbers of children. (For the 1970-forecast, numbers of marrying females by age group - observed and derived from forecast assumptions - should be multiplied by period intensities by age at marriage.) This is standardization for nuptiality.

Figure 5.4 indicates that marital fertility has had a greater contribution to errors in mean numbers of marital live birth per woman than nuptiality. The correspondence between the 1965-forecast and that of 1970 is striking. During the years with a strong decrease in fertility, errors are almost entirely explained by marital fertility rates. But after 1975, when fertility stabilizes, the contribution of nuptiality to forecast errors starts to gain importance, and around 1985 the error due to marital fertility becomes even smaller than that due to nuptiality.
Beex (1980, 65) ascribes forecast errors in nuptiality to unmarried cohabitation, which gained acceptance in the Netherlands in the 1970s. Indeed, the differences between observed percentages of married women for the year 1979 and corresponding percentages in the 1972-forecast (12 per cent overestimation for the age group 20-24, and 9 per cent overestimation for women aged 25-29), agree reasonably well with percentages of cohabiting women as observed in the Netherlands Fertility Survey of 1982 (16 per cent for age group 20-24, and 9 per cent for women aged 25-29) (NCBS, 1988, 46). Van de Mortel (1988, 37, 65) shows how the overestimation of the percentages of married women in the 1970-forecast, compared to actual percentages observed between 1970 and 1985, is caused by an overestimation of first marriage and remarriage, and an underestimation of divorce.
A last topic I want to take up in this subsection is a comparison between error patterns for absolute numbers of births, for birth rates and for mean numbers of children. On the basis of figure 5.5 one may conclude that the general pattern remains the same, whether one looks at absolute numbers, at crude rates, or at the mean number of marital live births per woman.

Figure 5.4. Mean number of marital live births per woman, the Netherlands

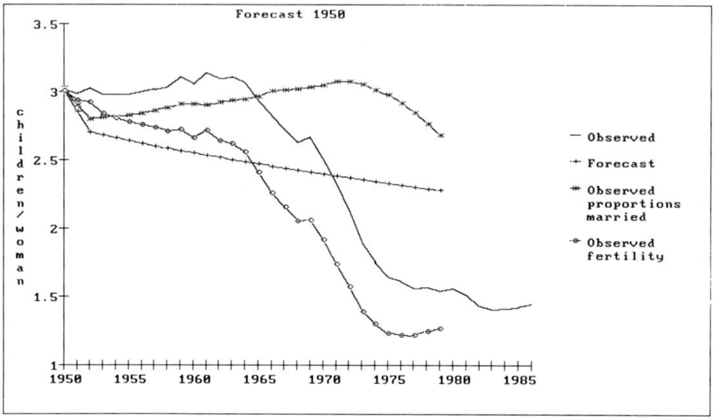

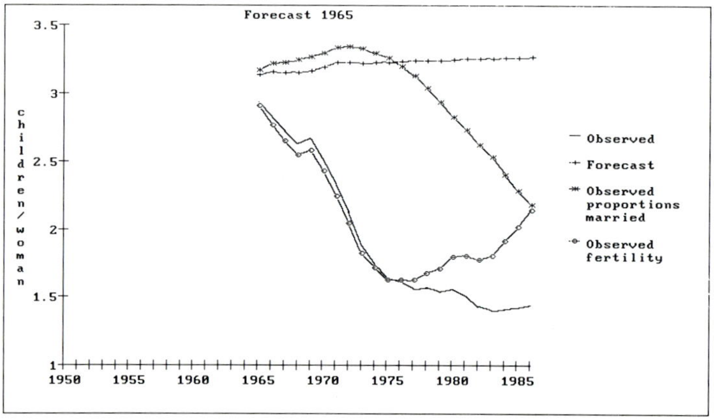

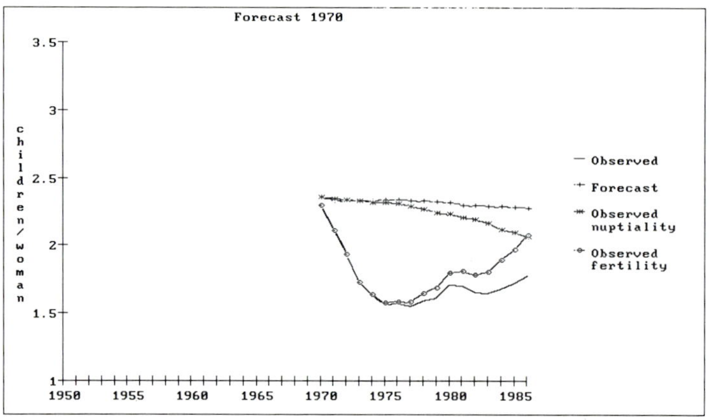

Figure 5.5. Percentage errors of fertility indicators, the Netherlands

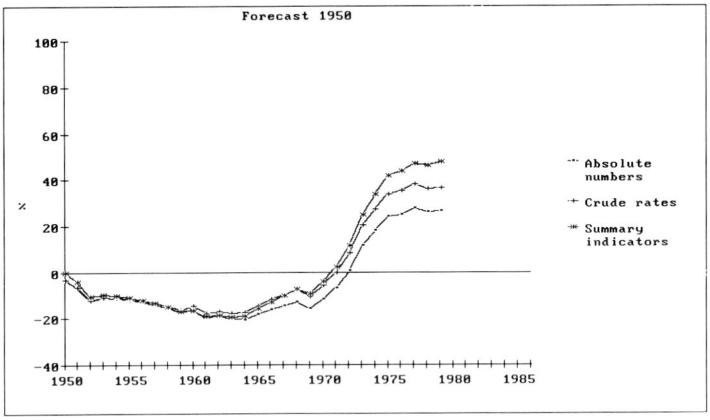

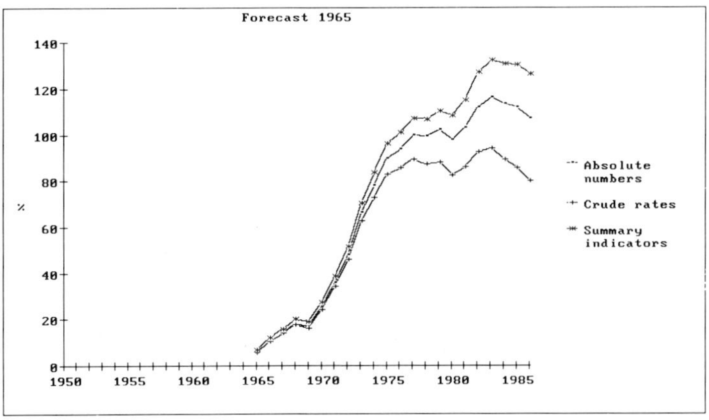

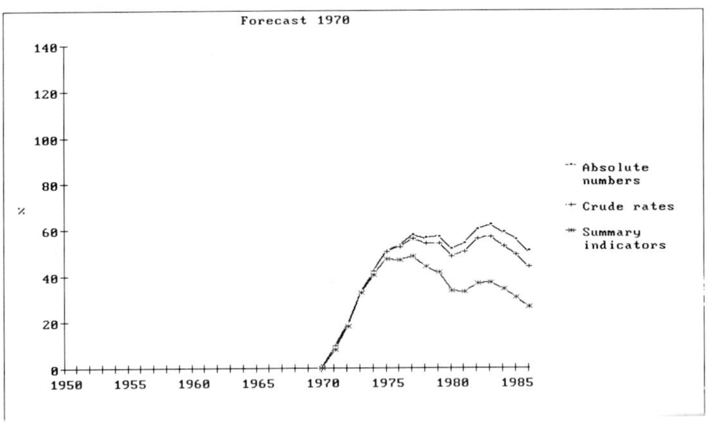

The curves are virtually indistinguishable for the first 5 (the 1970-forecast) to 15 (the 1950-forecast) years. After this initial period, sometimes absolute numbers display large errors compared to the other two indicators (the 1970-forecast), sometimes they show the smallest errors (the 1950-forecast). For the forecast of 1950 and that of 1970 the percentage errors in crude rates take a medium position; for the forecast of 1965 this is the case for absolute numbers. Because of these findings, any of the three indicators can be chosen for further analysis. Somewhat arbitrarily I selected the crude birth rate for the explanatory model of fertility forecast errors in chapter 7.

5.4.2. Mortality

For mortality, fluctuations in forecast accuracy also go hand in hand with changes in the observed trends. In the 1960s, male life expectancies dropped temporarily, which resulted in an accelerated growth of the annual numbers of deaths (figure 5.6). Since 1970, the decline in the life expectancy of males has come to a halt and the growth of the number of deaths has slowed down. The forecasts of 1965 and 1967 (in which the mortality assumptions were identical) were too optimistic for the first 5 to 10 years, resulting in an underestimation of mortality (see table 5.11). In the long term the average outcome was quite accurate, due to compensation effects after a period of 10 years (see also Van Poppel, 1984). The mortality forecasts of 1970, 1972 and 1975 were based partly on an extrapolation of the declining life expectancies of males experienced in the 1960s. This means that mortality was overestimated in those forecasts. It was not until the forecast of 1980 that forecasters had (almost perfectly) forecasted the improved mortality situation.

In its forecast publications, Statistics Sweden presents graphs of observed deaths, and deaths as computed in the forecasts of 1973, 1978, 1980, 1983 and 1986 (SCB, 1986, 77; SCB, 1989, 61). The similarity to the situation in the Netherlands, as shown in the lower panel of figure 5.6, is striking: actual deaths display an accelerating growth in annual numbers between 1960 and 1977, and next a steady growth at a lower level. But Swedish mortality forecasts produced between 1973 and 1980 implied a continuation of the trend which was observed until 1977. The new trend with more favourable mortality conditions, which set in in 1977, was only picked up by the forecasters in 1986. Thus Swedish forecasters exhibited the same assumption drag as their Dutch colleagues almost ten years earlier.

The summary indicator for which error patterns will be discussed next is the life expectancy at birth. For reasons of comparison I selected, as with fertility, the forecasts of 1950, 1965 and 1970, see table 5.12. Observed life expectancies at birth have been published by the Netherlands Central Bureau of Statistics in various life tables. In the forecast of 1950, during the entire forecast period, assumed life expectancies were equal to those observed in 1947-1949. Life expectancies at birth that were used in the

Figure 5.6. Deaths, forecasts and observations, the Netherlands

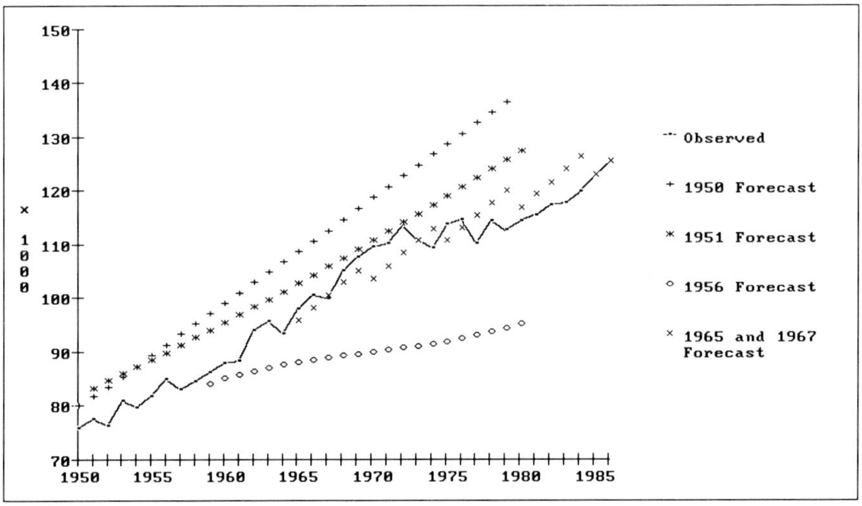

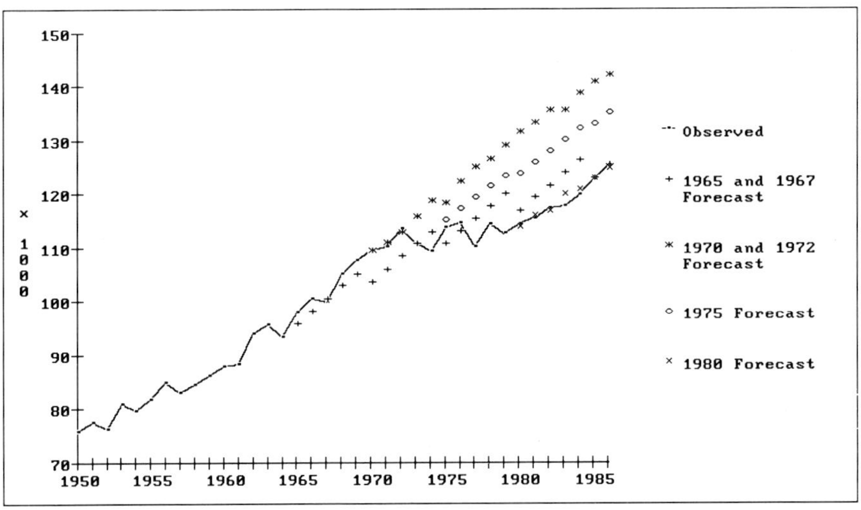

Table 5.11. Accuracy of mortality forecasts, by length of forecast period, the Netherlands

	Jump-off year of forecast													
	1950	1951	1956	1965	1967	1970	1972H	1972L	1975H	1975L	1980H	1980M	1980L	Total
						5 year period								
Number of deaths														
MPE (%)	7.1	8.1	-5.1	-1.6	-3.2	2.7	4.7	4.6	5.8	5.6	0.9	0.6	0.6	2.4
RMSPE (%)	7.4	8.6	5.9	2.0	3.8	4.5	5.4	5.3	6.7	6.5	1.2	1.0	1.0	5.2
U	0.9	4.4	1.7	0.4	1.2	2.0	2.5	2.5	2.7	2.7	0.7	0.5	0.5	1.5
Crude death rate														
MPE (%)	7.2	7.3	-5.1	-1.8	-2.8	2.9	5.0	4.9	5.8	5.8	0.7	0.4	0.4	2.3
RMSPE (%)	7.5	7.6	5.8	2.2	3.5	4.5	5.7	5.6	6.7	6.7	0.9	0.8	0.8	5.1
U	0.9	3.4	1.7	0.4	1.1	2.2	2.8	2.7	3.0	3.0	0.4	0.4	0.4	1.5
						10 year period								
Number of deaths														
MPE (%)	9.0	8.8	-8.0	-1.8	-2.5	6.4	9.1	8.8	7.8	7.6				4.5
RMSPE (%)	9.5	8.7	8.9	3.1	3.5	8.3	10.3	10.0	8.4	8.2				8.2
U	1.3	6.3	1.8	0.6	1.1	2.3	3.0	3.5	4.1	3.3				2.1
Crude death rate														
MPE (%)	9.5	7.9	-7.7	-2.8	-2.2	5.8	9.0	9.2	8.0	8.3				4.5
RMSPE (%)	10.1	8.1	8.6	3.6	3.3	7.4	10.1	10.5	8.6	9.0				8.3
U	1.3	4.7	1.9	0.6	1.1	2.7	3.2	3.3	4.5	4.7				2.0
						15 year period								
Number of deaths														
MPE (%)	10.0	7.5	-11.4	-0.5	-0.6	9.4	10.9	10.7						4.5
RMSPE (%)	10.5	8.1	12.8	3.5	3.5	11.1	11.9	11.6						9.8
U	1.8	2.1	1.8	0.7	1.2	2.6	3.9	4.7						2.2
Crude death rate														
MPE (%)	11.4	7.1	-11.0	-2.9	-0.9	8.0	10.3	11.0						4.1
RMSPE (%)	12.1	7.3	12.3	3.7	3.0	9.3	11.1	12.0						9.6
U	1.7	2.0	1.8	0.6	1.2	3.4	4.0	4.3						2.0

1965-forecast were calculated by Van Poppel (1984) on the basis of published age structures in 5-year age groups for males and females for the years 1965, 1970, 1975, ..., and numbers of live births over the period 1965-69, 1970-74, 1975-79 and 1980-84. Finally, information on life expectancies that were used in the 1970-forecast are to be found in the relevant publication (NCBS, 1971, 35).

Table 5.12 and the corresponding figure 5.7 clearly show the different trends for males and females, which were discussed earlier. The 1965 forecast contained increasing mortality rates for males aged 17-23 and 47-71. This was remarkable, compared to mortality forecasts made for other countries in that period. For instance, Pressat (1975, 55) notes that the statistical agencies of Sweden and the UK hesitated to make a similar pessimistic choice. In fact, Dutch forecasters were the first in the OECD to predict increasing male mortality in certain age classes (Pressat, 1975, 61).
In spite of the relatively good insights in future mortality trends which Dutch forecasters demonstrated in 1965, one notes an overestimation over the years 1965-1969 of the life expectancy for both sexes in the 1965-forecast. This must have been the reason for the rather pessimistic mortality assumptions of the forecast of 1970. In the early 1970s life expectancies at birth for males and females resumed the upward trend which they had displayed in the 1950s. Hence, the mortality assumptions of the 1970-forecast can now be considered a somewhat too strong reaction to the performance of those of the forecast of 1965. The NCBS, when analysing the accuracy of mortality of the 1965-forecast, notes that they had been too optimistic regarding male mortality rates for ages up to 54 years, and for females even up to 69 years (NCBS, 1971, 17). Hence, for the 1970-forecast, an extrapolation of age-specific mortality was made which implied an increase of male rates for the ages 10-19 (caused by increases in mortality from neoplasms and accidents) and 30-79 (neoplasms and cardiovascular diseases), and an increase of female rates for the ages 10-54 (accidents and cardiovascular diseases), see NCBS (1971, 33). As a result, the difference between males and females in the accuracy of the life expectancy of the forecast of 1970 is large: for males one notes an assumed decrease and for females only a slight improvement, whereas both sexes showed a considerable increase in reality.
The male RMSPE is higher than the female RMSPE for the forecasts of 1950 and 1965, see table 5.13. When one looks at the deviation from zero of the MPE the picture is similar. This agrees with the findings of errors in the age structure (section 5.3) which show that aged women are more difficult to forecast than aged men. The 1950-forecast has a Theil-coefficient of 1 since the assumed life expectancies were equal to those observed prior to the forecast, i.e. those from the life tables 1947-49.

The accuracy of the forecasts of life expectancy at birth in the Netherlands is somewhat different from that in the United States. Olshansky (1988, 494,

Table 5.12. Life expectancy at birth, forecasts and observed, the Netherlands

	Observed		Jump-off year of forecast					
			1950		1965		1970	
	M	F	M	F	M	F	M	F
1949	69.4*	71.5*						
1950	70.6**	72.9**	69.4	71.5				
1951	70.6**	72.9**						
1952	70.6**	72.9**						
1953	71.0***	73.9***						
1954	71.0***	73.9***						
1955	71.0***	73.9***						
1956	70.9	74.1						
1957	71.4	74.6						
1958	71.5	74.9						
1959	71.3	75.2						
1960	71.5	75.3						
1961	71.5	75.8						
1962	71.1	75.7						
1963	71.1	75.8						
1964	71.2	76.2						
1965	71.1	76.1			71.4	76.8		
1966	71.1	76.1						
1967	71.0	76.5						
1968	71.0	76.4						
1969	70.9	76.3			71.4	76.8		
1970	70.7	76.5			71.6	77.8	70.6	76.3
1971	71.0	76.7						
1972	70.8	76.8						
1973	71.3	77.3						
1974	71.6	77.6			71.6	77.8	70.6	76.3
1975	71.4	77.6			71.9	78.6	70.2	76.5
1976	71.5	78.0						
1977	72.0	78.4						
1978	71.9	78.5						
1979	72.4	78.9	69.4	71.5	71.9	78.6	70.2	76.5
1980	72.5	79.2			72.3	79.4	69.5	76.6
1981	72.7	79.3						
1982	72.7	79.4						
1983	72.9	79.6						
1984	73.0	79.7			72.3	79.4	69.5	76.6
1985	73.1	79.7					69.5	76.6
1986	73.1	79.6					69.5	76.6

* Life tables 1947-1949.
** Life tables 1950-1952.
*** Life tables 1953-1955.

Figure 5.7. Life expectancy at birth, forecasts and observations, the Netherlands

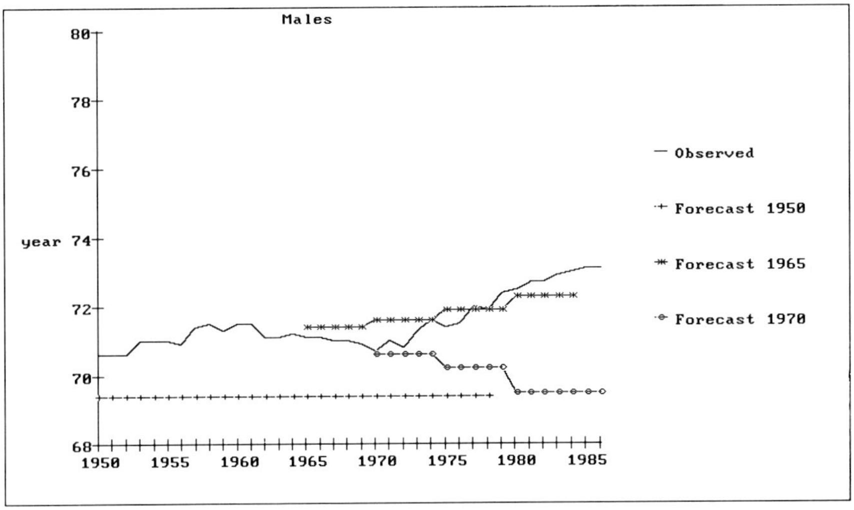

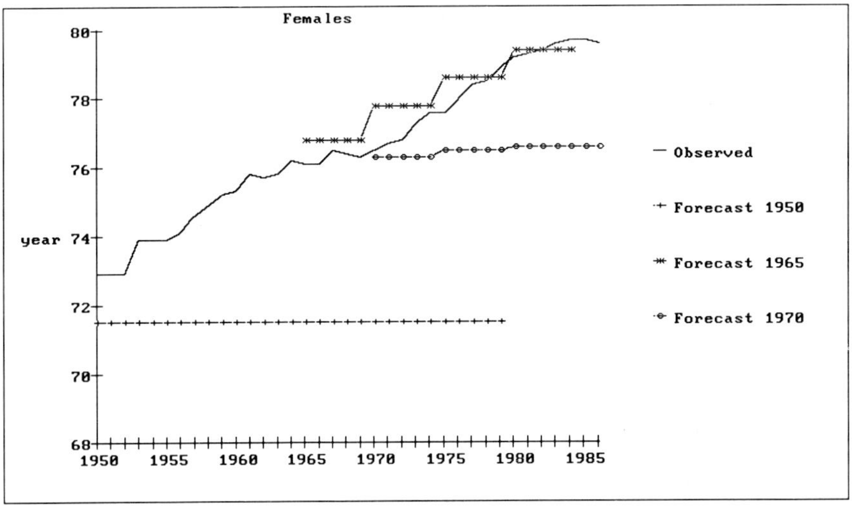

Table 5.13. Accuracy of life expectancy at birth, the Netherlands

	Jump-off year of forecast 1950				1965				1970			
	Length of forecast period 1-5 (years)	1-10	1-15	total	1-5 (years)	1-10	1-15	total	1-5 (years)	1-10	1-15	total
Males												
MPE (%)	-1.9	-2.2	-2.4	-2.5	0.5	0.6	0.4	0.2	-0.7	-1.5	-2.5	-2.8
RMSPE (%)	1.9	2.3	2.4	2.6	0.5	0.7	0.6	0.6	0.8	1.8	3.0	3.3
U	1.0	1.0	1.0	1.0	1.8	1.7	0.9	0.4	1.5	1.7	1.7	1.7
Females												
MPE (%)	-2.5	-3.3	-4.1	-5.7	0.7	0.9	0.7	0.5	-0.9	-1.6	-2.2	-2.4
RMSPE (%)	2.5	3.4	4.3	6.1	0.7	1.0	0.9	0.8	1.0	1.8	2.5	2.7
U	1.0	1.0-	1.0	1.0	3.1	1.2	0.5	0.3	1.0	0.9	0.9	0.9

495) presents observed US-life expectancies at birth for males and females for the period 1940-1986, and contrasts these figures with forecasts of life expectancies made up by the US Office of the Actuary since 1952. He notes that forecasts of life expectancy made in the period 1952-1977 did not follow exactly from the mortality trends that were observed in the years prior to the jump-off, and US-actuarians did not anticipate that life expectancy would continue to increase at such a rapid pace. Observed trends for males during the period from the mid-1950s to the end of the 1960s, show a temporary stagnation in a general upward trend. But, contrary to the case of the Netherlands, this did not result in forecasts of declining male life expectancies. The improvement of life chances was simply somewhat less in the US-forecasts of 1966 and 1974, than in earlier and in later forecasts.

A detailed international comparison was carried out by Preston (1974). He evaluates official mortality forecasts made around 1950 and 1960 in Australia, Canada, Japan, New Zealand and the United States. On the whole, Preston concludes, that for most groups defined by age and sex, mortality declined faster than was foreseen in the forecasts, in particular for females. This leads to large errors for females when mortality was assumed constant. Except for Japan, every forecast was over-optimistic for males in the age-range 40-65: future male mortality was underestimated for these ages in every one of the forecasts which Preston considered. His analysis reveals that resulting forecast errors are almost certainly attributable to unforeseen increases in mortality from neoplasms, cardiovascular diseases and motor vehicle accidents.

Finally, from the material which Vallin (1989) presents for official mortality forecasts produced in 1964, 1970, 1979 and 1985 for France, one may conclude that for France, too, female life expectancies were generally underestimated. However, contrary to Preston's findings for the five countries mentioned above, quite a number of French forecasts of male mortality (three out of six mortality variants) came out too low as well.

Figure 5.8. Percentage errors of mortality indicators, the Netherlands

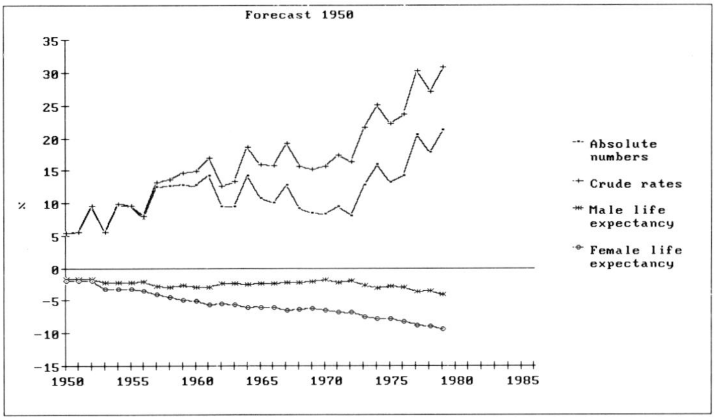

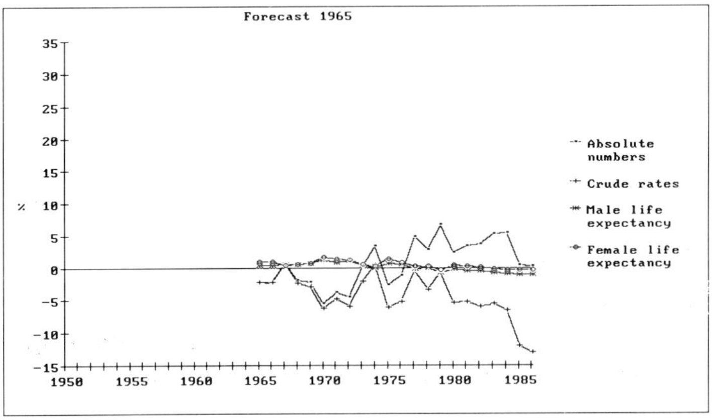

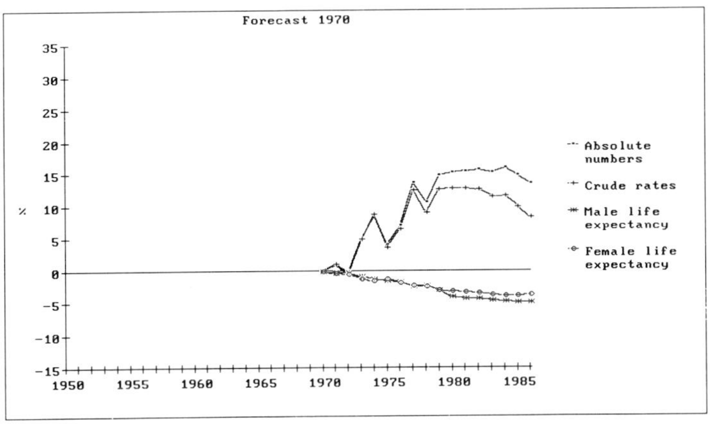

I now return to errors observed in the Netherlands and compare the performance of different indicators. Error patterns of life expectancies, death rates and absolute numbers of deaths show larger differences than those for mean numbers of children, birth rates and numbers of live births (cf. figure 5.8). This is, of course, partly due to the fact that an *improvement* of mortality (everything else remaining the same) results in an *increase* in life expectancy, but in a *decrease* of the death rate and the number of deaths. As the life expectancy enables one to analyse male and female error patterns separately, and possible distortions due to errors in age structures are to be avoided, I selected this summary indicator for further explanatory investigations in chapter 7.

5.4.3. External migration

External migration errors will be only briefly dealt with, since migration was not formally included in Dutch forecasts until 1975. Given the definition of a population forecast adopted in chapter 1 it will be clear that I shall assume that external migration "forecasts" made before 1975 were implicitly put at zero. An analysis by Schreurs (1989, 27) shows that Theil's U is slightly below 1 (0.92) for the 1975-based external migration forecasts which cover the period 1975-1985. The average U-value (over nine 10-year migration "forecasts" produced between 1950 and 1975) is 1.3, and the "forecasts" produced in the 1950s were better than average: 1.1 for the 1950-forecast, 0.6 for that of 1951, and 0.9 for that of 1956. Hence the formal introduction since 1975 of external migration as a component of change in Dutch forecasts generally improved their accuracy, although an assumption of zero net-immigration implicitly contained in the forecasts produced in the 1950s, did not yield less accurate results.
A brief analysis of "summary" indicators of external migration concludes

Table 5.14. Accuracy of emigration and immigration forecasts, the Netherlands

	Jump-off year of forecast				
	1975			1980[1)]	
	Length of forecast period			Length of forecast period	
	1-5	1-10 (years)	total (years)	1-5	total
Emigration					
MPE (%)	0.0	-4.9	-12.8	-2.9	-0.6
RMSPE (%)	2.1	8.4	29.9	5.0	5.0
U	0.5	1.8	4.6	0.9	0.8
Immigration					
MPE (%)	-9.0	-7.0	-14.7	9.3	2.8
RMSPE (%)	14.0	15.8	32.3	11.4	12.4
U	0.8	0.8	1.6	0.3	0.3

1) Medium variant.

Figure 5.9. Emigration and immigration, forecasts and observations, the Netherlands

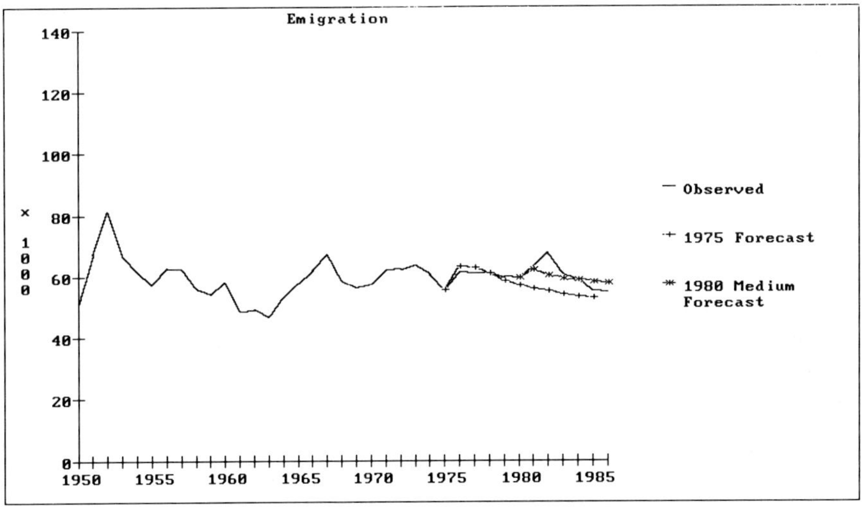

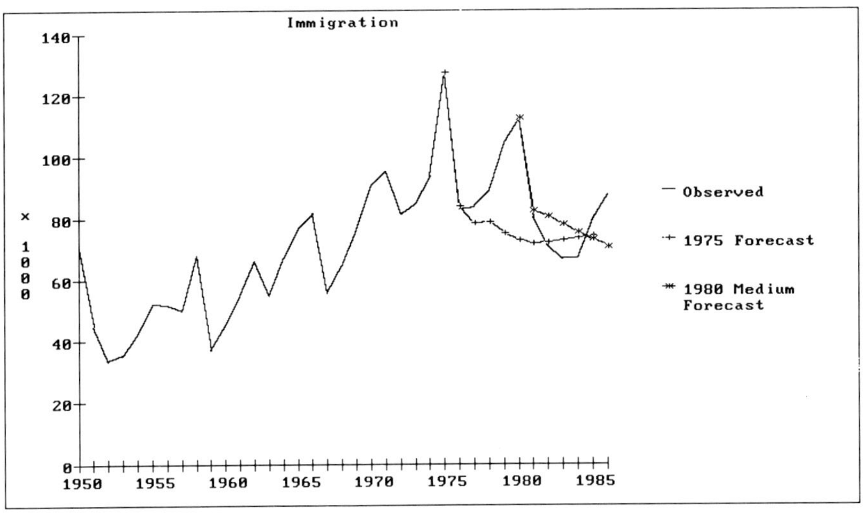

this section. A breakdown of net-migration into immigration and emigration is possible for the 1975-based forecast. With respect to the 1980-forecast, such a breakdown is, unfortunately, possible for the Medium variant only, not for the High variant or the Low variant. Observed and forecasted values of immigration and emigration are shown in figure 5.9. Figures for the 1975-based forecast can be found in the relevant publication (NCBS, 1976, 117), whereas information for the 1980-based forecast (medium variant) was obtained from archives of the NCBS. The error statistics in table 5.14 suggest that total numbers of emigrants are easier to forecast than total numbers of immigrants. This can be explained by the relatively stable course of emigration, which makes extrapolation less difficult than in the case of immigration. The low degree of accuracy of immigration forecasts is mostly due to the irregular flows of immigrants from non-European countries, e.g. Surinam, Turkey and Morocco.
Because of the high degree of stability of observed emigration, a naïve forecast would often be worth considering.

5.5 | Summary of findings

In this section I integrate the findings of the previous sections by answering three questions. Are recent forecasts more accurate than older ones? How fast do forecast errors grow as we look further into the future? And finally, how fast do those errors diminish as we approach a certain, initially chosen, year? The discussion will be limited to findings for absolute numbers and crude rates for fertility and mortality. Because of the limited number of forecasts which could be evaluated, the results that I obtained for summary indicators and for migration do not allow any firm answers to the questions asked in this section.

5.5.1. Are recent forecasts more accurate than older ones?
Table 5.8 suggests that the first turning point in forecasting methodology which, as was argued in chapter 4, took place between 1967 and 1970, has *not* resulted in smaller errors when forecasting fertility. The performance of the short-term (up to 5 years ahead) fertility forecasts of 1965 and 1967 was better than that of the 1970-forecast and the 1972-forecast. For medium-term forecasts those with jump-off years 1965 and 1967 are still better than that of 1970. Only for the long term (15 years) we see that the 1970-forecast has smaller errors than that of 1965, although the forecast of 1967 is still superior to these two. This is partly due to the actual course of fertility, as was argued in section 5.4.1. However, the problem we are facing here is that a comparison of the performance of subsequent forecasts will be hampered either by different observation periods (if we keep the length of the forecast period constant), or by different forecast horizons (if we consider the same observation periods). Thus, a firm conclusion cannot be reached here, and in chapter 6 I shall disentangle the effects of forecast period and observa-

tion period on forecast errors. But the general conclusion from table 5.8 is clear: errors do not decline monotonically if we look at the series of fertility forecasts.
A similar conclusion holds for the accuracy of mortality forecasts, which has fluctuated considerably since 1950 (table 5.11). The forecasts of 1965, 1967 and 1980 were exceptionally good; both for the absolute number of deaths and for the crude death rate, the MPE and the RMSPE are below average (except for the 5-year MPE of the 1967-forecast), and the U-values of the forecast of 1965 and that of 1980 are less than one.

5.5.2. How fast do errors increase as we look further into the future?
In section 5.4 I have analysed tables 5.8 and 5.11 horizontally. But we can answer the question at what rate errors increase as we look further ahead, by considering the columns in these tables.

The "total" column in table 5.8 shows that the MPE for all fertility forecasts taken together rises by approximately 7-8 percentage points every 5 years: 5% after 5 years, 12% after 10 years and 20% after 15 years. This linear trend is similar to that of the MPE for 0-4-year olds in figure 5.1. The RMSPE increases in a more or less linear fashion, too: 13% after 5 years, 24 to 25% after 10 years and 34 to 36% after 15 years. If we look separately at the different fertility forecasts, an exponential increase of the MPE can be seen in the 1965- and 1967-forecasts. The errors in the forecasts made in the 1970s level off after a period of 10 years. A linear increase of the RMSPE can also be found in the fertility forecasts of 1950 and 1965, whereas a levelling-off of the increase is found in the 1967-, 1970- and 1972-forecasts.
Figure 5.1 also seems to indicate that the error in mortality forecasts does not increase at as fast a rate as does the error in fertility forecasts. Table 5.11 supports this assumption. For all forecasts taken together, the MPE is 2% after 5 years, about double this amount after 10 years, and it more or less remains at this level after 15 years (4-5%). This levelling-off may be attributed in particular to the 1950-, 1951- and 1972-forecasts. For mortality, the RMSPE increases more slowly than at a linear rate, contrary to the RMSPE for fertility: 5% after 5 years, 8% after 10 years, and 10% after 15 years.
Thus, errors in mortality forecasts are not only smaller, but they also increase at a much slower rate than do the errors in fertility forecasts. This means that we can look further into the future for mortality than for fertility: the level of error reached after 5 years in fertility forecasts had not even been reached after 15 years for mortality.

5.5.3. How fast do errors diminish as we approach a certain fixed year?
In past centuries, the year 2000 had an almost magical ring to it. All sorts of future scenarios have focussed on this particular year. However, now that we are in the early 1990s, the importance of the year 2000 in forecast activi-

Table 5.15. Percentage errors of various forecast variables for the years 1975 and 1985, population forecasts since 1950, the Netherlands

	Jump-off year of forecast												
	1950	1951	1956	1965	1967	1970	1972H	1972L	1975H	1975L	1980H	1980M	1980L
1975													
Population size (1 Jan.)	-7.4	-1.1	0.4	3.7	0.1	0.4	-0.3	-0.3	-	-	-	-	-
Fertility													
Absolute number of births	23.9	44.3	44.3	89.9	47.3	50.7	32.4	22.1	-	-	-	-	-
Birth rate	33.7	45.9	43.7	83.1	47.3	50.1	32.8	22.5	-	-	-	-	-
Mortality													
Absolute number of deaths	13.2	4.7	-19.2	-2.7	-3.4	4.0	5.5	5.3	-	-	-	-	-
Death rate	22.3	5.9	-19.5	-6.2	-3.4	3.6	5.8	5.7	-	-	-	-	-
1985													
Population size (1 Jan.)	-	-	-	14.0	5.1	4.5	1.9	-0.0	0.4	-1.3	1.0	0.7	0.5
Fertility													
Absolute number of births	-	-	-	111.8	65.6	55.7	35.9	22.7	23.4	-3.0	13.4	8.9	4.4
Birth rate	-	-	-	85.8	57.5	49.0	33.4	22.8	22.9	-1.7	12.3	8.1	3.9
Mortality													
Absolute number of deaths	-	-	-	0.3	-0.4	14.8	13.9	13.7	9.0	8.5	0.2	0.2	0.2
Death rate	-	-	-	-12.0	-5.2	9.9	11.8	13.7	8.6	9.9	-0.8	-0.5	-0.2

ties has dwindled. We are increasingly venturing to look into the 21st century. One of the reasons is, of course, that uncertainty regarding forecast results decreases as the year 2000 comes closer. But how quickly does the uncertainty disappear?

It goes without saying that we cannot yet apply ex-post analyses to population forecasts for the year 2000. Instead, I have chosen the fixed years 1985 and 1975, and have looked at the forecast errors for demographic variables made in the successive population forecasts. In doing so I have eliminated the disadvantage of the analyses in the foregoing sections, namely that differences in accuracy could be the result of differences in the inherent uncertainty for population forecasts made in various decades since 1950.

Table 5.15 shows that neither for the population size on 1st January 1975 nor for the birth or death rates for that year, can one speak of a monotonously declining uncertainty as the forecast period becomes shorter. Who would have thought that the 1956-forecast would give an equally accurate estimation of the population size in 1975 as the 1970-forecast? And if we look at fertility, in 1975 this is what we get: an overestimation of 24% in 1950, an overestimation of 51% in 1970, and still as much as 22-32% in 1972! It looks as if the forecasters of the 1960s and the early 1970s were constantly misled by developments which took place in the years preceding the year in which the forecasts were made.

The ten-year period preceding 1975 had witnessed strongly declining birth rates. As from that year, the fertility level remained more or less constant. This might very well be the reason for the fact that the forecast errors for fertility (and thus for population size) for the year 1985 *did* decrease more or less monotonously.

6. A SEPARATION MODEL FOR THE ANALYSIS OF PERIOD, DURATION AND JUMP-OFF YEAR RELATED ERRORS OF POPULATION FORECASTS

In this chapter I describe and apply a model which disentangles the confounding effects of the observation period and the length of the forecast period on forecast errors in a series of forecasts. A formal separation of these two effects facilitates a much more profound analysis into the problem whether recent forecasts are more accurate than older ones, than could be given by means of the descriptive analysis in chapter 5.
Section 6.1 gives a discussion of the problem of confounding effects, and in section 6.2 I describe an additive (linear) separation model which builds on insights obtained earlier from age-period-cohort models (APC models). A solution to the identification problem inherent to models of this kind is proposed, and estimation aspects are discussed. Section 6.3 contains results of an application of the model to errors in fertility, mortality and international migration. For the first two components, errors in absolute numbers as well as crude rates are analysed. First I apply the model to the complete set of 203 observations for these two components (Forecasts 1950, 1951, 1956,..., 1980; calendar years 1951-1986), next a smaller subset of these observations (Forecasts 1975 and 1980; calendar years 1976-1986) is investigated. International migration is analysed in terms of errors in net numbers of immigrants for the smaller subset of observations only. Finally, I demonstrate in section 6.4 how the accounting equation of demographic growth can be used to relate model results for errors in numbers of births, deaths and immigrants, to results for errors in annual population growth.

6.1 | Problem formulation

In chapter 5 I presented a descriptive analysis of ex-post forecast errors in fertility, mortality and international migration, as they were observed in the

national population forecasts of the Netherlands which were produced between 1950 and 1980. An important issue which was taken up in that chapter was the following: are recent forecasts in the Netherlands more accurate than older ones, and, more particularly, were the changes in forecast methodology, noted in chapter 4 (one taking place between 1967 and 1970, and one around 1975), effective in the sense that forecast errors became smaller? In section 5.5.1 I argued that this question cannot be answered properly with a descriptive approach alone. Hence the answer provided in that section could only be a tentative one. The reason is that several effects are confounded in such a descriptive approach. Take, for example, the change between 1967 and 1970 from an age-specific period perspective to a marriage-duration cohort perspective (see section 4.2). One could compare the ex-post observed error of the 1967 forecast as exhibited in, say, the year 1972, to the ex-post error of the 1970 forecast for the same year. However, the 1967 forecast was five years old in 1972, and the 1970 forecast only two years. As forecast errors tend to grow with increasing length of the forecast period (see section 5.5.2), the errors of the two forecasts as observed in 1972 cannot be compared directly. On the other hand, when the disturbing effect of the length of the forecast period is held constant, one would be forced to compare ex-post errors which are observed in different calendar years. Therefore, when analysing the ex-post errors of a series of forecasts, one should carefully separate, for each forecast, the effects of forecast duration and calendar year.
Ahlburg (1982) concludes, after having investigated a number of live births forecasts produced by the U.S. Bureau of the Census (USBC) since 1953, that the introduction of a cohort perspective in 1964 for the projection of fertility has led to more accurate forecasts (compared to a naïve projection) than did the period methodology used earlier. Stoto (1983) analyses the error in US population growth for projections made by the Bureau of the Census between 1946 and 1971. He only considers the impact of jump-off year and duration on these errors, and he does not take into account possible confounding effects of calendar year. As Ahlburg and Stoto did not disentangle jump-off year, duration and period effects, their conclusions should be considered with great caution. Long (1987) analyses the rate of total population change in the US as forecast in USBC forecasts since 1945, and he suggests that some of the improvement in accuracy after 1970 may be due to methodological improvements in fertility analysis, but an alternative explanation he gives is the lower variability of fertility since the early 1970s. An isolation of period effects from forecast errors for fertility could tell us *how much* of the improved accuracy is simply due to smoother fertility developments.

In this chapter I propose to investigate ex-post errors in population forecasts by means of a model that distinguishes between period (calendar year) effects, duration (length of forecast period) effects and jump-off year effects. The jump-off year effect of a particular forecast would tell us what the "average" level of error is which is connected with that forecast. It can only be estimated accurately if confounding effects of period and of duration are removed from the observed forecast errors. By comparing jump-off year effects for subsequent forecasts we would be able to investigate a possible improvement of forecast quality over time.

6.2 | Model and estimation

The analysis of this chapter makes use of a separation model in which the ex-post error which corresponds to a certain forecast (i.e. to a certain jump-off year) and a certain year of observation of the forecast outcome is assumed to be composed of three terms: one inherent to that particular forecast, one corresponding to that particular year of observation, and one specific to the length of the forecast period (LOFP). This model is, in its technical appearance, equivalent to the well-known age-period-cohort model (APC model). However, the model in this chapter deviates from current APC models in its field of application, and hence in its interpretation of the estimated effects. Most APC models are applied to time series of aggregate data (for such diverse issues as fertility, mortality, earnings, crime and religion) broken down by age and (birth) cohort of the individuals involved. Condon and Pullum (1988) give an application of APC models to underestimation of population counts in the United States Censuses between 1880 and 1980. As they analyse the difference between estimated and real value of the census population, this comes close to my application of APC models to forecast errors in a series of forecasts. This new application enables me to evaluate the quality and the accuracy of historical population forecasts.

The *cohort effect* of the usual APC model is interpreted here as the contribution to the forecast error of the particular forecast (jump-off year). This *forecast effect* is seen as an indicator which expresses in one number the various aspects which are specific for the forecast under consideration. The forecast effect involves not only technical aspects, such as the model which was used to prepare the forecast, or the approach which was used to analyse and extrapolate the demographic components. It also includes elements of a wider context, for instance organisational, social and political aspects (see chapter 3 for a discussion of the impact which these issues may have on forecasting). Because one forecast effect will be estimated for each forecast produced by the NCBS since 1950, and since these forecasts were made at irregular intervals, forecast effects will be non-periodical. This is different from the usual approach in demographic APC models, where cohort effects are estimated for subsequent one-year or five-year cohorts.
The *age effect* used in APC models reflects in the present analysis the error contribution of the LOFP. This so-called *duration effect* expresses the increase in uncertainty as the forecast period grows longer. The longer the forecast period, the greater the probability that external factors, which are

assumed constant in the forecast, will change.
The *period effect* of the present model corresponds to the period effect of APC models. It quantifies a set of factors that express social and demographic conditions observed during a certain period in time, and their impact on forecast errors, independent of the circumstances under which the forecast was produced (these are summarized in the forecast effect) or the length of the forecast period (captured by the duration effect).

The separation model is used to analyse ex-post observed errors in population size, numbers of births, crude birth rates, numbers of deaths, crude death rates, and net-numbers of immigrants contained in national population forecasts produced by the NCBS since 1950. Its most simple version takes on the following general form.

$$E(i,j,k) = F(i) + P(j) + D(k) + u(i,j,k) \qquad (6.1)$$

Expression (6.1) explains the error (in population size, births and deaths - number or crude rate - or immigrants) for the forecast with jump-off year i, as observed in calendar year (or period) j, for an LOFP of k years, as a sum of a forecast effect F(i), a period effect P(j), a duration effect D(k) and a residual u(i,j,k). In the model it is assumed that all errors which are observed for one particular forecast, e.g. the one with jump-off year 1967, have the same forecast effect (to be denoted by F(1967) in this example), irrespective of the LOFP or the year for which these errors are observed. Similarly, all errors observed in a certain calendar year have the same period effect, regardless the jump-off year of their respective forecasts, while the duration effect for an LOFP of k years applies to all errors of forecasts which are k years "old". The residual u describes the part of the forecast error that is not explained by the forecast effect, the period effect and the duration effect.

Model (6.1) contains *main effects* only. It can be extended with *interaction effects,* when necessary. For instance, when the forecast with jump-off year i is assumed to exhibit period effects for period j that are different from those in forecasts with other jump-off years, an interaction term FP(i,j) may be added to model (6.1). Interaction effects will turn out to be necessary in a few cases.

Model (6.1) is an *additive* or *linear* model. The choice for such a model, as opposed to a *multiplicative* (or *log-linear*) model which is generally used in APC analyses, is based on the fact that forecast errors E are expressed in this chapter as the *difference* between forecast values (absolute numbers or crude rates) and corresponding observed values. Thus the dependent variable may take on positive as well as negative values (it is a ratio variable in the case of errors in crude rates). An overestimation of fertility in, say, the 1965-based forecast yields positive error values. Negative values for the dependent variable to be explained by the separation model result from an underestimation of the demographic variable. The advantage of an additive model is that it can adequately handle situations with both positive and negative values of the dependent variable. (A multiplicative model would of course have been appropriate if the forecast errors would had been expressed as the *ratio* of forecast and observed values. But taking logs on both sides of such a multiplicative model would have resulted in a model of the form expressed above.) If the dependent variable were a cell count or a cell probability the additive model would have been less attractive than the multiplicative model (Bishop et al., 1975, 24).

Estimation of the parameters of model (6.1) is not straightforward, due to the *identification problem.* This problem is caused by the fact that the three parameters (forecast's jump-off year, observation year and duration) are linearly dependent: duration (expressed in years) equals calendar year of observation minus forecast jump-off year. Hence it can be seen as perfect multicollinearity.
Several solutions to the identification problem have been proposed in the literature on APC modelling. The volume edited by Mason and Fienberg (1985) provides an extensive discussion of the problem. It summarizes other studies which have dealt in detail with the identification problem, for instance Glenn (1976), and Fienberg and Mason (1978). It also contains a chapter by Hobcraft, Menken and Preston who review APC models in demography. Further main references are Clogg (1982), Rodgers (1982) and Hagenaars (1985).

To solve the identification problem, three major types of approach exist (Willekens and Baydar, 1984). The first two approaches focus on the effect parameters (independent variables), the third deals with the dependent variable.
The first solution is to add an extra specification for one or more effects, for instance, to assume that the first period effect equals the second one, or that the first cohort effect is equal to the last cohort effect. However, such additional restrictions have a strong impact on the estimates of the effects of age, period and cohort, see, for instance, Willekens and Baydar (1984, 34), and Hagenaars (1985, 361). Therefore, they should be introduced with great care, and always be guided by theory or other a priori considerations.
The second way of overcoming the identification problem is to replace either one of the age, period, or cohort variables by an underlying variable which is not linearly dependent on the other two variables. For instance, one could parametrize one or more of the three variables, see, for instance, Hagenaars (1985, 267, 294).
Willekens and Baydar (1984) suggest and apply a third approach which is particularly useful in the analysis of time series of age-specific data for fertility, mortality and other demographic phenomena. They propose to split

each observed demographic rate (being defined in the Lexis diagram either over a square or a parallelogram) into its two constituent triangles. In this way they create twice as many observations on the dependent variable, each observation being uniquely defined for a certain age, a certain period, and a certain cohort. This breaks the perfect relation between the three independent variables, because persons who experience the relevant event are assigned to non-overlapping cohorts.

The identification problem shall be dealt with in the present analysis in two ways. First, by introducing an appropriate parameterisation of the duration effects. The descriptive analyses in section 5.5.2 have shown that forecast errors broken down by length of forecast period follow a regular, largely increasing, pattern. In the following sections I shall report on some experiments with a linear, a square-root, and an exponential form of the duration dependence. Second, by imposing restrictions on the effects to be estimated. For instance, in the case of errors in forecasts of live births, I shall assume in some models that there are only three forecast effects: one for the forecasts with jump-off years between 1950 and 1967, one for the forecasts of 1970, 1972 and 1975, and one for the forecasts produced after 1975. This assumption on equal forecast effects parallels the transition in forecast methodology which took place between 1967 and 1970, and the one which occurred around 1975. In all models I apply a restriction on the period effects: although annual data are available, for reasons of parsimony I estimate period effects for quinquennial periods only (except when single-year period effects differ too much within one five-year period). Thus, within one period of five years, the restriction is that the five period effects are all equal.

Special attention will be given to the issue of *forecast variants*. The NCBS calculated a high and a low variant for the forecasts of 1972 and 1975, and since 1980, three variants are computed: a high variant, a medium variant and a low variant (see section 4.3). This means, in the interpretation of cohort effects used in this chapter, that there are *two* cohorts 1972, *two* cohorts 1975, and *three* cohorts 1980. Therefore I included an additional term in expression (6.1), which can be interpreted as *variant effect*. An analogous approach was taken by Willekens and Baydar (1984), who introduced a region effect in their APC analysis of time series of age-specific fertility rates for the Netherlands.

Expression (6.1) can now be reformulated as follows.

$$E(i,j,k,\ell) = F(i) + P(j) + D.k + V(\ell) + u(i,j,k,\ell) \qquad (6.2a)$$

for the linear specification of duration dependence,

$$E(i,j,k,\ell) = F(i) + P(j) + D\sqrt{k} + V(\ell) + u(i,j,k,\ell) \qquad (6.2b)$$

for the square-root specification of duration dependence, and

$$E(i,j,k,\ell) = F(i) + P(j) + B\{1\text{-}\exp(-D.k)\} + V(\ell) + u(i,j,k,\ell) \quad (6.2c)$$

for the exponential specification of duration dependence.
Here the variable ℓ is used to denote the forecast variant (high, medium or low) for which the ex-post error is observed, and $V(\ell)$ is the effect of variant ℓ. The coefficient D in expression (6.2a) reflects the slope of the straight line that describes duration effects, and in expression (6.2b) it stands for the "slope" of the square-root function. In expression (6.2c) the coefficient B is expected to be positive, because the duration effect $B\{1\text{-}\exp(-Dk)\}$ is likely to rise as k increases.

Estimation of the model sketched above was done by Ordinary Least Squares (OLS) Regression. The approach taken here thus deviates from the standard approach in APC models, where estimation is frequently carried out by means of Maximum Likelihood, starting from either a multinomial or a Poisson distribution which is assumed to underlie the data. In OLS one assumes a Normal Distribution (at least for interval estimates).
Estimation of multiplicative APC models may be analysed within the framework of log-linear models. Hagenaars (1985) discusses the advantages and the disadvantages of a regression approach in log-linear analysis. Regression should be avoided if a Normal Distribution underlying the data were an unrealistic assumption, if impossible predictions would result from the model, or if the results of such an analysis would be difficult to interpret (Hagenaars, 1985, 12-14). None of these objections against a regression approach apply here.
Instead of minimizing the sum of squared residuals, I could have minimized the χ^2-statistic as a measure of the distance between observed values and expected values. Bishop et al. (1975, 348-357) and Reynolds (1977, 187-213) discuss the equivalence of this approach and weighted least squares estimation when the dependent variable represents cell probabilities in a cross table.

For the purpose of estimation, dummy variables in vector form were introduced for each effect on each level. For instance, the errors in live births were represented as a column vector with 203 elements. The first 29 elements in this vector apply to the 1950 forecast (with data for the years 1951-1979), the next 29 elements apply to the 1951 forecast (for the years 1952-1980) and so on. Nine dummy variables were introduced for the forecast effects, one for each jump-off year. Each dummy variable is a column vector of size 203, with ones at appropriate places, and zeros elsewhere. The dummy variable linked to the 1950 forecast contains 29 ones as its first 29 elements, the remaining 184 elements being zero. The 1951 forecast dummy variable has a one in positions 30-58, etc. Dummy variables for the period effects were constructed in a similar manner. The matrix which con-

sists of the entire set of dummy vectors is known in APC analyses and related research as the *design matrix* (Reynolds, 1977, 198).
The design matrix approach makes it very easy to introduce interaction effects and equality constraints on effects. For instance, a possible interaction between the forecast effect of the 1965 forecast and the calendar year 1969 would require one extra dummy (vector) variable in the design matrix, which is simply found as the product of the 1965 forecast dummy variable and the 1969 period dummy variable. And the condition that the 1951 period effect be equal to the 1952 period effect is dealt with by replacing the 1951 period dummy variable and the 1952 period dummy variable by a new one (the 1951-1952 period dummy variable) which is obtained as the sum of these two.

It should be noted, that the usual disclaimer in APC analysis applies here as well: estimated effects as such are without meaning. It is only the *difference* between two effects which can be interpreted.

For reasons of convenience, I estimated the model without a constant, and dummies for all forecast effects were introduced in the model. In case a constant ("overall effect") is included, all estimates are relative to this constant, and results are similar to those obtained without a constant. In order to avoid estimation problems, the period effect for the most recent period was fixed to zero, i.e. P(1986)=0. (The approach, frequently applied in the analysis of cross-tables, in which the sum of the period effects and the sum of the forecast effects both are set equal to zero, was not followed here.)

6.3 | Results

There are 203 observations on errors in live births, deaths and migrants, and 217 observations on errors in total population size. (The 14 additional observations for stock variables result from the fact that, for each of the 14 forecasts, I have one extra observation on population size, viz. the one as of December 31st 1986, compare tables 5.2-5.5.) I start the analyses with tables 6.1 and 6.2, which summarize the observed errors. Model results are presented in tables 6.3-6.8.
Table 6.1 contains a number of summary statistics for errors in annual births and deaths, both in absolute numbers (expressed in thousands) and in crude rates (per thousand), errors in absolute numbers of annual net-immigrations, and errors in annual population growth. The upper panel contains information on observed errors for the period 1951-1986, as "produced" by the forecasts 1950-1980. The period of the mid-1960s until the mid-1970s was a turbulent one in the demographic history of the Netherlands: fertility declined sharply, and death rates increased temporarily, especially for males (see also sections 5.4.1 and 5.4.2). Therefore the analyses in this chapter will also be carried out for the "smooth" period 1976-1986, to test the

robustness of the conclusions for the choice of a particular period. Summary statistics for the period 1976-1986 are shown in the lower panel of table 6.1. They apply only to the forecasts of 1975 and 1980.

Table 6.1. Summary statistics for variables describing forecast errors in annual births, annual deaths, annual net-immigration, and annual population growth

	Mean	Standard deviation	Coefficient of variation
Period 1951-1986, forecasts 1950-1980			
Errors in live births			
- absolute numbers (in thousands)	39.81	57.02	1.43
- crude birth rates (per thousand)	2.70	3.71	1.37
Errors in deaths			
- absolute numbers (in thousands)	5.08	9.44	1.86
- crude death rates (per thousand)	0.37	0.80	2.19
Errors in net-immigration (in thousands)	-16.81	22.17	-1.32
Errors in annual population growth (in thousands)	20.08	53.30	2.65
Period 1976-1986, forecasts 1975 and 1980			
Errors in live births			
- absolute numbers (in thousands)	9.65	15.61	1.62
- crude birth rates (per thousand)	0.67	1.04	1.55
Errors in deaths			
- absolute numbers (in thousands)	5.75	5.00	0.87
- crude death rates (per thousand)	0.41	0.39	0.96
Errors in net-immigration (in thousands)	-3.13	16.81	-5.37
Errors in annual population growth (in thousands)	4.40	27.72	6.30

The statistics in table 6.1 confirm the results of the descriptive analysis in chapter 5. The mean error (taken over all 203 observations) in absolute numbers of live births exceeds those in numbers of deaths and immigrations. The mean error in annual population growth is lower than that in births because of compensating effects of errors in deaths and migrations. The population growth error is not simply the sum of errors in births, (minus that of) deaths and migrations, because administrative corrections (see section 5.1 on data checks) are not taken into account here. The standard deviations of errors in mortality over the period 1951-1986 are relatively small, compared with those for births and migrations, as demonstrated by the coefficient of variation. For the shorter period 1976-1986 the variation in migration errors is extremely high, while that in mortality errors is very small (coefficients of variation with values of -5.37 and 0.87 or 0.96, respectively).

Table 6.2 contains a further breakdown of the mean errors in the upper panel of table 6.1 by jump-off year, by period and by duration. It can thus be used as a yardstick against which estimated effects in the following sections can be measured. Therefore, table 6.2 will not be reviewed here, except for the fact that the duration breakdown provides a good illustration of the usefulness of disentangling confounding effects for period, duration and jump-off year. For example, the errors in crude birth rates display a regular rise from 0.25 to 1.63 for durations between one and six years. The rise continues for durations 7-11, but at a much higher level: from 2.33 to 2.89. Further jumps occur after durations of 11, 14, 16 and 19 years. The latter two jumps imply a sudden shift downwards of the trend. These jumps can be explained by the fact that I have observations for six calendar years (and hence for six durations) for all nine forecasts 1950-1980, but that there are observations for 11 durations for eight forecasts only (forecasts 1950-1975), and observations for 14 durations for seven forecasts, etc. The upward shift between durations of six and seven years is caused by the fact that the forecast of 1980, of which the average error of 0.87 per thousand is much lower than the overall error of 2.70 per thousand (see table 6.1), is no longer contained in durations of seven years or more. Similar explanations can be given for jumps occurring at other durations.

6.3.1. Errors in live births

Table 6.3 shows a selection of the results that were obtained for absolute numbers of births for the period 1951-1986. Effects are expressed in thousands.

Model 1 is a model which contains only forecast effects, duration effects (in parametrized form according to the square-root function) and period effects. In this simple model, forecast effects of forecasts made between 1950 and 1967 are made equal, and similarly for the forecasts of 1970 and 1972, and those of 1975 and 1980. This enables me to test the assumption that the two changes in forecasting approach which took place between 1967 and 1970, and around 1975 (see chapter 4), went together with an increase in accuracy. The estimates demonstrate that, indeed, this seems to be the case. A forecast which was produced between 1950 and 1967 contains an error

Table 6.2. Mean errors for fertility and mortality, by jump-off year, period and duration

	Fertility		Mortality	
	Absolute numbers (in thousands)	Rates (per thousand)	Absolute numbers (in thousands)	Rates (per thousand)
Jump-off years				
1950	-15.68	-0.78	11.54	1.32
1951	18.65	1.35	6.36	0.54
1956	25.93	1.80	-14.90	-1.14
1965	136.19	8.50	0.87	-0.38
1967	74.44	5.03	0.19	-0.11
1970	86.18	5.82	12.30	0.71
1972	62.23	4.32	13.60	0.90
1975	6.64	0.51	9.02	0.73
1980	13.34	0.87	0.77	0.02
Periods				
1951-55	-12.09	-1.21	6.52	0.60
1956-60	-19.86	-1.76	7.58	0.69
1961-65	-29.10	-2.25	3.28	0.38
1966-70	-1.28	0.16	-1.53	0.03
1971-75	54.42	4.21	0.75	0.16
1976-80	70.25	4.94	7.19	0.50
1981-85	58.78	3.69	8.70	0.45
Durations				
1	3.62	0.25	1.63	0.14
2	8.74	0.59	3.24	0.26
3	16.72	1.17	2.94	0.22
4	18.28	1.26	4.02	0.30
5	21.00	1.46	3.65	0.26
6	23.52	1.63	3.67	0.26
7	33.27	2.33	6.39	0.48
8	36.89	2.56	7.65	0.56
9	40.40	2.78	8.53	0.61
10	43.29	2.90	7.60	0.52
11	43.84	2.89	7.32	0.48
12	54.03	3.51	7.46	0.45
13	55.08	3.53	6.36	0.36
14	58.22	3.72	7.52	0.42
15	66.97	4.24	3.49	0.12
16	71.41	4.50	3.39	0.10
17	70.84	4.45	0.58	-0.06
18	72.28	4.46	-0.04	-0.14
19	72.06	4.35	-0.94	-0.24
20	68.37	4.19	-1.62	-0.21
21	74.50	4.58	-1.98	-0.25

which is 65 thousand (=202710-137350) births higher than that of a forecast produced between 1970 and 1972, irrespective of the period in which the error is observed. Also, the forecast effect of the 1970 and the 1972 forecasts is no less than 84 thousand births higher than that of the 1975 forecast and the 1980 forecast.

Table 6.3. Regression results for absolute numbers of births, 1951-1986

Effects	Model 1	Model 2	Model 3
		in thousands	
Jump-off years			
1950	202.71	22.19	14.56
1951	202.71	54.42	46.83
1956	202.71	58.15	50.20
1965	202.71	102.16	95.40
1967	202.71	60.39	53.85
1970	137.35	80.75	72.98
1972	137.35	39.33	31.69
1975	52.88	-19.76	-24.89
1980	52.88	3.26*)	-3.03
Periods			
1951-55	-190.29	-50.63	-49.11
1956-60	-182.21	-65.21	-65.71
1961-65	-182.78	-81.79	-82.55
1966-70	-156.73	-71.86	-71.96
1971-75	-82.22		
1976-80	-32.28	10.41	10.42
1981-85	5.21*)	9.04	8.74
1971		-54.70	-53.39
1972		-38.21	-37.06
1973		-19.53	-17.87
Durations			
k		0.73	
$\sqrt{k}$	-15.18		5.03
Variants			
High		11.90	11.90
Low		-11.90	-11.90
Interactions			
LPF65		3.97	3.88
LPF67		1.87	1.75
LPF75		2.20	1.84
R^2(adj.)	0.76	0.99	0.99

Note: all estimates are significant at the 5 % level, except those denoted with an asterisk.

The coefficient of $\sqrt{k}$ shows that the estimated impact of LOFP on errors in numbers of births is somewhat smaller than the estimates of forecast effects: the value of -15.18 implies that a forecast period of, say, four years would result in an additional error due to LOFP of $-15.18\sqrt{4}$ or approximately -30 thousand births (underestimation), and after 16 years one could expect an underestimation of some 60 thousand births.
The value of R^2 (adjusted for the number of independent variables and the number of observations) indicates that the fit can be much improved. An analysis of the residuals of model 1 revealed that the forecast effects of the forecasts of 1950, 1965 and 1967 should be estimated separately from those of the forecasts of 1951 and 1956. Similarly, forecasts 1970 and 1972 should have distinct effects. Main effects for the variants should be included, as well as an interaction effect for LOFP and the forecasts of 1965, 1967 and 1975: the errors of these forecasts tend to increase more sharply than on average, as the length of the forecast period grows.
Model 2 shows how the fit improves when these factors are taken into account. More important, however, is the fact that any drop in forecast effect between 1967 and 1970 is absent in model 2; the forecast effect of the 1970 forecast is even 20 thousand births *higher* than that of the preceding forecast. (The difference is highly significant; its estimated standard deviation is only 4.0 thousand.) What remains is the forecast effect of the 1975 forecast, which is 60 thousand births lower than that of the 1972 forecast. Again, the difference is significant at the 5% level.
The interactions between LOFP and the effects of the forecasts of 1965, 1967 and 1975 were introduced as follows: a dummy variable was defined as the product of the independent variable k (with values 1, 2, 3 etc. for durations of one, two three etc. years) and the dummy variables related to the forecast effect of these forecasts. Hence, each of these three forecasts was given its own k, so to speak.
The estimates of the interaction effects LPF65, LPF67 and LPF75 in model 2 indicate that the forecast error increased much more sharply in these three forecasts than in general: when we consider the rate of increase for these forecasts, the values of 3.97, 1.87 and 2.20, respectively, *should be added* to the general value of 0.73, which applies to *each* forecast. Thus, due to duration effects (main effect *plus* interaction effect), the error in the 1965 forecast after 10 years, say, would be $10 * (0.73 + 3.97) = 47$ thousand births.
Introducing a high variant in the forecasts has led to an additional error of nearly 12 thousand births, compared to the situation with only one variant (or compared to the medium variant). The dummy variable for the Medium Variant was left out in order to make estimation possible; forecasts containing only one variant were denoted as "Medium", and those with two variants as "High" and "Low".

A linear form of the effect of LOFP (see expression 6.2a) performs as well as a square-root form (expression 6.2b), as is shown in the estimation results of model 3. However, both perform much better than an exponential form.

The exponential form works well for the forecasts of 1950, 1951,..., 1972. The estimated value of B in a model similar to models 2 and 3, but with exponential duration dependence (see expression (6.2c)) and D fixed at a value of 0.5, is -16.85, and its t-value is -4.46 (results not shown here). Patterns of estimates for period effects and forecast effects correspond closely to those of models 2 and 3. However, the fit of the exponential duration model is very bad for the forecasts of 1975 and 1980. A further experiment with *both* B and D regarded as parameters, using non-linear estimation, did not result in estimates for these two parameters which were significantly different from zero. The conclusion is that an exponential form of duration dependence works well in times of relatively large errors which increase or decrease more or less uniformly, but not when demographic trends are fairly stable.

In many of the models which were estimated (not all are shown here), the Durbin-Watson statistic indicated positive autocorrelation in the residuals. Methods which deal with this problem usually transform dependent variables E(i,j,k) to E'(i,j,k) = E(i,j,k)-ρE(i,j-1,k), where ρ is the correlation between successive residuals. Independent variables are transformed analogously. This approach was not pursued here, because transformed variables would yield estimates which are difficult to interpret. The consequence is that coefficient estimators are still unbiased, but not efficient; estimators for variances and covariances are such that t-statistics tend to be exaggerated (Maddala, 1983, 283). In the explanatory analysis of chapter 7 due account will be given of possible autocorrelations in model residuals.

Forecasters do not formulate their fertility assumptions in terms of absolute numbers of live births - they usually describe future fertility by means of the period total fertility rate, or by its cohort counterpart, i.e. the completed fertility rate. Information on such summary measures from subsequent NCBS population forecasts is incomplete, as was argued in section 5.4.1. Therefore, a comprehensive evaluation of such measures using the approach of this chapter is not possible. However, as a first approximation to the period total fertility rate I computed the crude birth rate for each year in each forecast, and results of the separation analysis for errors in these crude birth rates are given in table 6.4. Estimated coefficients are expressed per thousand.

The results for models 1-4 demonstrate that the general patterns of forecast effects and period effects are similar to those observed in table 6.3 for errors in absolute numbers. A remarkable difference, however, is the absence of an effect due to the length of the forecast period, at least in parametrized form. Except for model 1 of table 6.4 (for which the fit is rather poor), no model with a significant duration effect could be fitted. A model (not shown in table 6.4) which consisted of forecast effects, period effects and LOFP-effects in unparametrized form as given in expression 6.1 (with durations of

Table 6.4. Regression results for crude birth rates, 1951-1986

Effects	Model 1	Model 2	Model 3	Model 4
		per thousand		
Jump-off years				
1950	11.65	3.13	3.57	4.06
1951	11.65	5.05	5.48	5.97
1956	11.65	4.86	5.30	5.71
1965	11.65	8.78	9.87	9.47
1967	11.65	5.00	5.94	5.67
1970	7.85	5.57	5.83	6.14
1972	7.85	2.88	3.10	3.41
1975	2.61	-1.52	-1.15	-1.09
1980	2.61	0.21*)	0.29*)	0.54
Periods				
1951-55	-11.57	-5.38	-5.82	-6.12
1956-60	-11.29	-6.23	-6.67	-6.84
1961-65	-11.32	-6.98	-7.42	-7.50
1966-70	-9.00	-5.66	-6.25	-6.11
1971-75	-3.87			
1976-80	-1.10	0.75	0.45	0.49
1981-85	0.02*)	0.53	0.44	0.40
1971		-3.89	-4.38	-4.29
1972		-2.57	-3.04	-2.95
1973		-1.23	-1.62	-1.60
1974		-0.78	-0.75	
Durations				
k				
√k	-0.80	0.12*)	0.12*)	
Variants				
High		0.78	0.60	0.60
Low		-0.78	-0.60	-0.60
Interactions				
LPF65		0.07		0.07
LPF67		0.06		0.06
LPF75		0.20		
LPF75H			0.25	0.29
LPF75L			0.08*)	0.11
R^2(adj.)	0.79	0.98	0.98	0.98

Note: all estimates are significant at the 5 % level, except those denoted with an asterisk.

1, 2, 3,..., 15 years) was estimated (with one period effect for the years 1951-1970, and one forecast effect for the forecasts of 1950, 1951 and 1956, to circumvent the identification problem). None of the duration effects were significantly different from zero (t-values were between -1.37 and 0.37). The pattern was that of a regular rise for durations 1-5 years, and very irregular and negative estimates for durations 6-15 years. My interpretation is that LOFP-effects in table 6.3 are created by the strong overestimation of fertility in the forecasts produced between 1965 and 1972, which has caused sharply rising numbers of births in these forecasts, and hence an increase in total population size. Indeed, projected birth rates in the forecast of 1965 show an initial increase from 22 per thousand to 24 per thousand after ten years, but after this period the trend is slowly downwards to some 23 births per thousand inhabitants. On the other hand, the absolute number of births in this forecast display a steady rise, see figure 5.3. In other words, the increase in total population was steeper than that in errors in the number of births. Similar patterns can be observed for the forecasts of 1967 and 1970. This explains why LOFP-effects disappear when one evaluates crude rates instead of absolute numbers, i.e. when a correction is made for the size of the population.

Model 1 takes on a very simple form, with main effects only (and an independent variable to take LOFP into account). The effect of the 1970 forecast (an overestimation of the birth rate with 7.85 per thousand) is significantly lower than that of the 1967 forecast (11.65 per thousand) - the estimate of the standard error of this difference is 0.44. Model 2 is similar to model 3 in table 6.3. Among others, separate duration effects were introduced for the forecasts of 1965, 1967 and 1975. The coefficient for the (square-root) duration effect is not significantly different from zero.
An analysis of the residuals of models 1 and 2 revealed that both the high and the low variant of the 1975 forecast should have their own LOFP-effect. Therefore, I included two interaction effects LPF75H and LPF75L, by means of the product of the forecast 1975 dummy, the variants dummy, and the independent variable k.
The results for model 3 show that the interaction between variant (high or low) and LOFP for the 1975 forecast is significant, and that both effects have the expected sign: positive for the overestimations in the high variant, and negative for the underestimations in the low variant. The fact that such an interaction effect should be taken into account for the 1975 forecast, and not for the 1972 forecast (which contains a high and a low variant) or the 1980 forecast (with a high, a medium and a low variant) can be explained as follows. As shown in figure 5.2, the 1975 forecast has a margin between the high and the low variant which widens much more rapidly in the first few years of the forecast period, than is the case for the forecasts of 1972 and 1980. This can be considered as a reaction of the NCBS forecasters to the large errors in the forecasts of 1965-1972, in which the sudden drop in fertility that took place in the Netherlands between 1964 and 1975 was insuffi-

ciently anticipated. The margin between the high and the low variant of the 1980 forecast was rather narrow, thus reflecting the optimism among the NCBS forecasters about the predictability of births after a period of stable fertility trends since 1975. This illustrates nicely the general phenomenon that, after a steady movement of demographic trends in a consistent direction, confidence increases, and the range of admitted future possibilities narrows. When a sudden change creates uncertainty the margin between high and low is made very large (Keyfitz, 1977, 235).
Attempts to fit an adequate model with linear or exponential duration effects were unsuccessful, for the reasons described above. Therefore I fitted model 4 to the data. No main effects for duration dependence of errors in crude birth rates are included - duration effects are captured by means of interactions with selected forecast effects (1965, 1967 and 1975 turned out to be necessary) and variant effects.
The general conclusion (which became also clear from models 1-3) is that starting with the 1965 forecast, a steady improvement of forecast accuracy can be noted: forecast effects decline regularly from a high 9.47 per thousand for the 1965 forecast, to a low 3.41 per thousand for the 1972 forecast, and further to a level of -1.09 per thousand for the 1975 forecast. The 1980 forecast has an effect of only 0.54 per thousand. However, the difference between forecast effects of 1967 and 1970 is such (the former effect is 0.47 per thousand *higher* than the latter effect, and the estimated standard deviation of this difference is no less than 2.86 per thousand), that one cannot conclude that the change in forecast methodology between 1967 and 1970 has led to smaller forecasting errors, at least not immediately. But it should be noted that the forecasts of 1975 and 1980 have forecast effects for (errors in) birth rates which are significantly lower than those of the preceding forecasts. This supports the assumption of an improvement of forecast accuracy for fertility due to the changes in the forecast approach which took place around 1975.

Joop de Beer has pointed out to me that the interaction between forecast effects and period effects may be important. One may assume that the 1975-forecast is more accurate than earlier ones mainly because actual fertility levels in the first few years of that forecast differed only slightly from those immediately prior to the jump-off year 1975. I think this point is relevant, and I estimated a model similar to model 4, but with an additional interaction term for the forecast of 1975 and the period 1976-1980. The estimation results (not shown here) indicate that, indeed, a weak interaction effect is present: the estimated coefficient of the interaction term is -0.765 per thousand, with a t-value of -1.757. Moreover, contrary to model 4, the forecast effect connected to the 1975 forecast does not significantly differ from zero (the estimate is -0.15 per thousand, its t-value is -0.23). More important is, however, the fact that the difference between the effects of the forecasts of 1972 and 1975 in this model (although smaller than in model 4, in which it is 4.5 per thousand, see table 6.4) still amounts to 3.6 per thousand, with an estimated standard error of this difference equal to 0.8 per thousand. Thus the relatively stable fertility observed between 1976 and 1980 explains about 20 per cent of the improvement in average accuracy between 1972 and 1975.

Similarly, one could argue that the steep fertility decline in the first half of the 1970s had a negative effect on the accuracy of the 1970 forecast. Without such a decline, that forecast would perhaps have been better than the 1967 forecast, which was produced with a less advanced method. I estimated a model similar to model 4, but with an additional dummy variable for the interaction between an effect of the 1970-forecast and an effect of the period 1971-1975. The results are very close to those of model 4, *including* the 1970-forecast effect: 6.07 in the model with interaction, and 6.14 in model 4. The interaction effect itself is estimated to be 0.24 per thousand; its t-value of 0.85 indicates that it is not significant. Therefore, the steep decline in fertility immediately after 1970 has had virtually no *additional* impact on the accuracy of the 1970-forecast, compared to the effect which that decline had on *all* forecasts made between 1950 and 1970.

6.3.2. Errors in deaths

A first attempt to separate the effects of period, forecast's jump-off year and LOFP in errors in death forecasts is given in table 6.5 under "Absolute numbers, model 1". This model contains main effects for periods and jump-off years, and a parametrized effect for LOFP. I did not include a main effect for variants, because for each of the nine forecasts produced between 1950 and 1980, mortality assumptions were formulated in one variant only. The differences between variants in errors in deaths forecasts which are shown in table 5.4 are thus purely a consequence of differences of population at risk, due to fertility variants for the forecasts of 1972, 1975 and 1980, and migration variants for the 1980 forecast.

Conclusions regarding forecast effects for errors in absolute numbers of deaths can only be drawn for the forecasts of 1956, 1970, 1972 and 1980. Estimated standard errors for the effects of the remaining forecasts are so large that the estimated effects in this model cannot be very reliable.
A careful analysis of the residuals of model 1 and some experimentation led to the conclusion that interactions between LOFP and the forecasts of 1950, 1956 and 1970 should be included in the model. Moreover, the residuals for the years between 1971 and 1985 were so large, and, at the same time highly irregular, that period effects should be expressed for single calendar years. Hence model 2 was estimated, which reveals significant effects for the forecasts of 1956, 1970, 1972, 1975 and 1980 (in addition, the estimate for the 1967 is almost significant at the 5% level, as its t-value of -1.98 is slightly above the critical value for these 179 degrees of freedom; the estimated standard error of the difference between the effects of the forecasts of

Table 6.5. Regression results for absolute numbers of deaths and death rates, 1951-1986.

Effects	Absolute numbers Model 1 (in thousands)	Absolute numbers Model 2 (in thousands)	Crude rates Model 3 (per thousand)	Crude rates Model 4 (per thousand)
Jump-off years				
1950	2.88*)	-0.81*)	0.83*)	0.43
1951	-2.25*)	2.03*)	0.04*)	
1956	-22.67	-3.95	-1.67	
1965	-5.43*)	-1.09*)	-0.84	
1967	-5.93*)	-1.73*)	-0.56*)	
1970	6.43	6.46	0.28*)	0.39
1972	7.41	10.97	0.51*)	0.47
1975	3.25*)	6.94	0.28*)	0.31
1980	-5.19	1.78	-0.36*)	-0.03*)
Periods				
1951-55	2.77*)	2.62	-0.01*)	
1956-60	4.17*)	1.60	0.23*)	
1961-65	4.32*)		0.38*)	
1966-70	-0.87*)	-4.92	0.20*)	
1971-75	-2.96*)		0.07*)	
1976-80	1.45*)		0.28*)	
1981-85	2.93		0.27	
1969			-0.25	
1970			-0.37	
1971		-7.10	-0.27	
1972		-9.00	-0.38	
1973		-5.38		
1974		-2.33		
1975		-6.53	-0.17	
1976		-5.39		-0.09*)
1977			0.32	
1978		-2.10		
1979			0.31	
1980			0.19	
1981			0.14	
1983			0.15	
1985			-1.89	
Durations				
$\sqrt{k}$	1.95	1.59	0.08*)	
Interactions				
LPF50		0.53		0.06
LPF51			-0.04	
SQRLPF51			0.30	
LPF56		-0.99	-0.08	
SQRLPF65			-0.13	
LPF70		0.46	0.04	
LPF72H			0.05	
LPF72L			0.06	
LPF75			0.05	
R^2(adj.)	0.87	0.96	0.85	0.97

Note: all estimates are significant at the 5 % level, except those denoted with an asterisk.

1967 and 1970 is only 1.11 thousand deaths, whereas the difference itself is 4.73 thousand deaths). The pattern is that of rising forecast effects for the forecasts produced between 1967 and 1972, and a decline thereafter.
As was pointed out in section 5.4.2, male life expectancy dropped temporarily in the Netherlands in the 1960s, resulting in an accelerated growth of the annual number of deaths. Since 1970, this decline in the life expectancy of males has come to a halt and the growth of the number of deaths has slowed down. The forecasts of 1965 and 1967 (in which the mortality assumptions were identical, see their estimated effects in model 2) were too optimistic for the first five to ten years. The mortality forecasts of 1970, 1972 and 1975 were based for a part on an extrapolation of the declining life expectancy of males as observed in the 1960s. This means that mortality was overestimated in those forecasts, and this is in agreement with the fact that the estimates for F70, F72 and F75 are significantly higher than that for F67.
Note that in 1956 (main effect *plus* interaction between LOFP and forecast effect = -4.9!) the NCBS forecasters took a too positive view with respect to life chances.
In both models the forecast effect attached to the 1975 forecast is significantly lower than that of the 1972 forecast. This suggests that the shift in forecast approach which was introduced around 1975 went together with an improvement of the accuracy of the mortality forecasts, as was the case with the fertility forecasts discussed in section 6.3.1.
Do errors in mortality grow faster or slower than those in fertility? A comparison of the estimated k-coefficient of model 2 in table 6.3 (0.73) to that of $\sqrt{k}$ in model 2 of table 6.5 (1.59) reveals that, after approximately 5 years ($1.59^2/0.73^2=4.74$), errors in deaths are lower *and* grow slower than those in births. On the other hand, a comparison of the two $\sqrt{k}$-coefficients (5.03 for model 3 in table 6.3 and 1.59 for model 2 in table 6.5) leads to the conclusion that, starting with a duration of one year already, errors in deaths grow about three times slower (5.03/1.59=3.16) than errors in births. Residuals of the models involved in these two conclusions explain why they are partly contradictory. Duration patterns in table 6.2 suggest that one can be much more confident about the latter conclusion than about the former.

Results of a separation analysis applied to errors in crude death rates are shown in table 6.5, models 3 and 4. As these rates are expressed per thousand, so are the estimated effects.
Model 3 contains main effects only. In comparison to the same model applied to errors in absolute numbers (model 1), less forecast effects are significantly different from zero. Yet the pattern of differences between forecast effects for subsequent forecasts is very similar to that of models 1 and 2. Period effects are all insignificant, except for that of the first half of the 1980s.
An analysis of the pattern of the residuals of model 3 revealed that a number of interactions would be necessary in order to improve the fit, in partic-

ular interactions with LOFP. Moreover it would be advisable to include period effects for single calendar years in the period 1970-1985. The results are shown under "model 4" in table 6.5.
The most important conclusion is that, similar to the case of fertility in the previous section, any main effect due to the length of the forecast period disappears once errors in rates are analysed instead of errors in absolute numbers. However, in the case of mortality, duration effects are still present in the form of interactions with forecast effects. The 1951 forecast, for example, has a strong duration effect, shown by LPF51 and SQRLPF51. (The latter interaction effect was defined as the coefficient of the product of $\sqrt{k}$ and F51.) Errors in death rates thus grow according to $0.30\sqrt{k}-0.04k$ in this forecast, except for period effects between 1969 and 1983 (here k stands for the length of the forecast period). After they have attained their maximum value (which occurs for k=14 years, in other words in 1965) period effects start to dominate a few years later.
LOFP-effects show relatively strong differences between forecasts, including a change of sign: the forecasts of 1956 and 1965 have *negative* duration effects, while the duration effects for the other forecasts (1950, 1951, 1970, 1972 and 1975) are positive. These strongly fluctuating duration effects between forecasts explain why a common duration effect for all forecasts could not be fitted.
The forecast effect of the 1975 forecast is significantly larger than that of the 1980 forecast and, at the same time, significantly smaller than that of the 1972 forecast. This again supports the assumption that the change in forecast approach which took place around 1975 brought about at least some improvement of forecast accuracy.

6.3.3. Errors in international migration
International migration was not taken into account until the forecast of 1975, at least not systematically. Therefore, a comparison between observed migration and forecast migration can only be carried out for the forecasts of 1975 and 1980, over the period 1976-1986. Of course, if the usual interpretation of the notion of a forecast is followed, being the most plausible estimate of future demographic trends, it could be argued that international migration was assumed (at least implicitly) to be absent in the early forecasts, see also section 5.4.3. But an analysis along the lines of this chapter of migration errors for the whole period 1951-1986 would then lead to forecast errors equal to (minus) observed net-immigrations for the forecasts of 1950, 1951, ... , 1972. This is somewhat artificial, as it would imply period effects only for the years 1951-1974, and no forecast effects or duration effects. Therefore I applied the separation model only to migration errors of the 1975 and the 1980 forecast, observed for the years 1976-1986.
Model 1 of table 6.6 contains main effects for the jump-off years, the periods and the variants, as well as a parametrized duration effect. Only the duration effect and the effect for the period 1981-1985 are significantly dif-

ferent from zero. Results for model 2 show how the fit can be improved when period effects for distinct years are taken into consideration. The estimated period effects for model 2 reflect the volatile character of international migration (see also figure 5.9). The results show clearly that errors in migration forecasts are dominated by period effects: these effects are much larger than the forecast effects. The strongly negative effect which was estimated for the 1980 forecast is a consequence of the underestimation of migration during the mid-1980s, see also table 5.5 and figure 5.9.

Table 6.6. Regression results for absolute numbers of net-immigrations, 1976-1986

Effects	Model 1	Model 2
	in thousands	
Jump-off years		
1975	23.59*)	1.69*)
1980	9.66*)	-9.10
Periods		
1976-80	-16.42*)	
1981-85	22.07	
1977		-4.11*)
1978		-4.82*)
1979		-19.82
1980		-26.37
1981		13.69
1982		30.59
1983		30.32
1984		30.07
1985		15.57
Durations		
k		-2.51
$\sqrt{k}$	-14.34	
Variants		
High	2.12*)	
Low	-2.12*)	
R^2(adj.)	0.67	0.91

Note: all estimates are significant at the 5 % level, except those denoted with an asterisk.

6.3.4 Errors in births and deaths for the forecasts of 1975 and 1980

Three main conclusions can be drawn from the analyses of errors in fertility and mortality in sections 6.3.2 and 6.3.3:

- fertility forecasts exhibit a substantial improvement in accuracy around 1975, while any improvement between 1967 and 1970 is absent;
- duration effects due to length of the forecast period disappear as soon as errors in crude rates are considered, rather than errors in absolute numbers;
- period effects and forecast effects are of approximate the same order of magnitude.

The third conclusion may be clarified as follows.
Table 6.3 indicates that forecast effects for errors in numbers of births (models 2 and 3) are between -20 (thousand) and + 100 (thousand). Period effects fluctuate between -72 and + 10. The order of magnitude of these two types of effects is approximately the same. The same can be noted for errors in birth rates (forecast effects between -1.5 and +9.9, period effects between -7.5 and 0.8, see models 2-4 in table 6.4), in numbers of deaths (forecast effects vary between -4.0 and +11.0, and period effects between -9.0 and +2.6, see model 2 in table 6.5) and in death rates (forecast effects between 0.0 and 0.5, period effects between -0.4 and +0.3, see model 4 in table 6.5). Although estimated effects as such cannot be interpretated but only differences between estimates, the fact that effects are of the same order of magnitude implies that, generally speaking, their differences are, too.

The three general conclusions given above apply to a period of turbulent demographic developments in the Netherlands: fertility declined sharply between 1964 and 1975, and death rates increased temporarily at the end of the 1960s, especially for males (see also sections 5.4.1 and 5.4.2). Therefore some of the analyses of the preceding sections will be repeated here for the "smooth" period 1976-1986, to test the robustness of the last two conclusions formulated above for the choice of a particular period[1].

The results in table 6.7 show that the last conclusion also holds for the shorter time period: for fertility as well as for mortality, for absolute numbers as well as for crude rates, forecast effects are not substantially larger or smaller than period effects. However, the finding for the period 1951-1986 that duration effects are only present in errors in absolute numbers of births and deaths and not in errors in corresponding crude rates, does not hold for the period 1976-1986. In spite of several experiments with the form of the duration dependence, I have been unable to fit any model for errors in absolute numbers of births which includes a main effect for LOFP in parametrized form. For mortality, the situation is the opposite of that for fertility: duration effects, although not very strong, are very significant. For instance, the estimated standard error of the coefficient of 0.04 in the model of errors in crude death rates is only 0.006.

An explanation for the fact that duration effects in fertility errors, when controlling for population size, disappear for the period 1951-1986 (i.e. are present for absolute numbers, but not for crude rates), but are even absent for absolute numbers for the period 1976-1986 (forecasts of 1975 and 1980 only), was given implicitly in the discussion of table 6.4: the forecasts produced between 1965 and 1970 were responsible for the presence of duration effects in long-term errors in numbers of births. These forecasts are not considered in the short-term analysis.
For mortality, the fact that a main effect due to LOFP shows up in errors in

Table 6.7. Regression results for fertility and mortality, 1976-1986

	Fertility		Mortality	
Effects	Absolute numbers (in thousands)	Rates (per thousand)	Absolute numbers (in thousands)	Rates (per thousand)
Jump-off years				
1975	-2.04*)	-0.15*)	3.44	0.28
1980	8.65	0.51	-4.08	-0.35
Periods				
1981-85	3.46	0.28		
1977	0.34*)	0.03*)	4.79	0.32
1978	-4.36*)	-0.30	2.05	0.13*)
1979	-4.74	-0.31	5.39	0.38
1980	-10.56	-0.69	3.41	0.27
1981	-6.65	-0.37		
1982	4.21	0.39		
1983	8.99	0.70		
1984	7.14	0.54		
1985	5.30	0.38		
Durations				
k	0.44*)	0.03*)	0.56	0.04
Variants				
High	4.73	0.33		
Low	-4.73	-0.33		
Interactions				
LPF75H	2.88	0.20		
LPF75L	-0.85*)	-0.04*)		
R^2(adj.)	0.98	0.98	0.96	0.95

Note: all estimates are significant at the 5 % level, except those denoted with an asterisk.

death rates observed for the forecasts of 1975 and 1980, and *not* in errors for all forecasts can be explained by looking once more at table 6.5. Duration effects are only present through interaction with forecast effects. Because these interactions are negative for the forecasts of 1956 and 1965, and positive for the forecasts of 1950, 1951, 1970, 1972 and 1975, a general duration effect could not be identified. When analysing errors in death rates for the shorter period, the duration pattern for the 1980 forecast turns out to be very similar to that of the 1975 forecast, and a main effect due to LOFP is present. Indeed, the estimate of 0.04 for the coefficient of k in table 6.7 is very close to the estimates in table 6.5 for the interactions between duration and forecast effects for the forecasts of 1970, 1972 and 1975.
A final observation concerns the relative magnitude of the forecast effects, when demographic components are compared. Tables 6.6 and 6.7 demonstrate that, when looking at absolute numbers for fertility, mortality and migration for the period 1976-1986, the forecast effects are roughly of the same magnitude: their absolute values range from 1.7 to 9.1. However, period effects for fertility and mortality are much smaller than those for migration.

6.4. | The link between estimates for births, deaths, migrations, and annual population growth

In this chapter, errors in absolute numbers of births, deaths and migrations were taken as the simple difference between forecast values and observed values. These errors were analysed by means of a linear model, using Ordinary Least Squares estimation. Now, simple demographic accounting tells us that annual population growth is the sum of births minus deaths plus migration surplus. These three facts together imply that if we estimate period, duration and forecast effects for errors in annual population growth, these estimates are related to corresponding estimates (assuming the same model) for errors in births, deaths and migrations. It can be easily shown that, when investigating errors in annual population growth, any OLS-estimate for a particular effect equals the sum of the corresponding effect for errors in births minus that for deaths plus that for net-immigration, provided one works with the same regressors and observations. (I am grateful to Evert van Imhoff for pointing this out to me.)
More generally, let b be the OLS-estimate of coefficients in a linear regression model with independent variables x and a dependent variable y. Furthermore, let y_i (i=1,...,n) be a set of dependent variables such that y is a linear combination of the y_i ($\Sigma m_i . y_i = y$), with b_i the corresponding estimates, and m_i a set of constants. Then simple calculation shows that the estimates b and b_i (i=1,...,n) are related as $\Sigma m_i . b_i = b$.
To demonstrate this, I calculated errors in annual population growth from table 5.2, and fitted the model in table 6.8 shown under "Population growth" to these errors. Next, the same model (with a period effect of 1986

fixed to zero) was fitted to errors in absolute numbers of births, deaths and net-migrations.

Table 6.8. Regression results for absolute numbers of births, deaths, net-immigrations and annual population growth, 1951-1986

Effects	Population growth	Births	Deaths	Net-immigrations
	in thousands			
Jump-off years				
1950	110.09	-63.67	-0.78*)	168.92
1951	144.16	-29.22*)	-5.76*)	163.47
1956	152.55	-12.55*)	-24.94	135.52
1965	150.30	58.29	-3.43*)	83.59
1967	118.00	35.03	-6.18*)	71.66
1970	97.24	44.12	6.70	54.51
1972	51.09	10.58*)	7.96	42.88
1975	13.04*)	-30.60	4.23	42.06
1980	14.21*)	-10.12*)	-3.79*)	14.64
Periods				
1951-55	-110.13	26.91*)	8.11*)	-125.10
1956-60	-131.32	-0.71*)	9.18*)	-113.61
1961-65	-148.24	-29.91	8.36*)	-107.69
1966-70	-120.84	-35.44	1.37*)	-79.87
1971-75	-70.28	5.48*)	-1.47*)	-73.20
1976-80	-24.68	23.43	2.25*)	-42.38
1981-85	17.24	14.00	3.35	9.08
Durations				
k	-2.81	3.37	0.51	-5.73
Variants				
High	12.85*)	11.90	0.17*)	1.16*)
Low	-13.02*)	-11.90*)	-0.17*)	-1.16*)
Interactions				
LPF65	3.75	3.55	-0.24*)	-0.03*)
R^2(adj.)	0.92	0.96	0.87	0.75

Note: all estimates are significant at the 5 % level, except those denoted with an asterisk.

The results in table 6.8 show that the relationship described above only holds approximately. For instance, the estimated effect for the 1950 forecast in the model which describes errors in population growth is 110.09. The sum of the corresponding effects in models for errors in births, deaths and net-migration is -63.67-(-0.78)+168.92=106.03, or some four thousand too low. In general, estimates for period effects and forecast effects in the population growth model are four to six thousand persons higher or lower than the linear combination of the corresponding effects for the three demographic components. The reason is that the demographic accounting equation, although being valid in the forecast model, does not hold exactly for observed data, due to administrative corrections (see section 5.1). But within the bounds defined by administrative corrections (in general an annual net-effect of a few thousand persons), the expected relationship between estimated effects is indeed present in table 6.8.

Note

1) There are three dimensions in the data analysed in this chapter: calender year, jump-off year, and duration. The robustness of the conclusions could, in principle, be tested by applying the model to parts of the data in each of these three directions. I chose the jump-off year dimension, because this perspective is most relevant for present forecasts. If one is interested in short-term forecasting, it is natural to divide the data set along the duration dimension, for instance by restricting observations to the first five durations of each forecast.

7. AN EXPLANATORY MODEL FOR EX-POST OBSERVED FORECAST ERRORS

In the two preceding chapters I described and analysed the ex-post forecast errors for the Netherlands, which were observed for the national population forecasts produced by the NCBS between 1950 and 1980. The purpose of the present chapter is to explore possible links between those errors and potential explanatory factors. This will be done in the form of an explanatory model.

Explanatory factors were identified in section 3.4, partly on the basis of theoretical considerations, but also based on empirical findings.

In section 7.1 I discuss the models to be used and the dependent and independent variables. To enhance interpretation, separate models are formulated for the absolute value of the error and its sign. The former type of models are linear, and they are estimated with Ordinary Least Squares. The latter are Probit models and these are estimated with Maximum Likelihood techniques. Section 7.2 presents results for errors in crude birth rates, in the life expectancy of males and females, and in immigration and emigration. In section 7.3 I select several models which are used to generate out-of-sample "predictions" (in the econometric sense) of forecast (in the demographic sense) errors for the period 1984-2000.

The idea to link forecast errors to covariates in an explanatory model is not new. Makridakis and Hibon (1979) developed regression equations expressing the accuracy of time-series methods as a function of factors such as randomness, seasonality, trend-cycle and the number of data points describing the series.

7.1 | Selection of model variables

7.1.1. Dependent variables

Several models are described in this chapter. For fertility, there are two dependent variables. First, the absolute value of the forecast error in the crude birth rate is taken as the dependent variable. Next, the sign of the

error is modeled. The reason to model the amplitude of the error and its sign separately is the following.
Difficulties in interpretation would arise if the dependent variable would be defined as the observed forecast error including its sign, i.e. the simple difference between forecast value and observed value. To see this, consider the range of values which this error may attain. A strongly negative error indicates a very inaccurate forecast. As the error value grows, it becomes less negative reflecting an improvement of accuracy. However, as it grows further, as soon as it exceeds the zero level, accuracy starts to decrease again. Because of the non-uniform behaviour of the error defined this way, the absolute value of the difference between observation and forecast was taken as the dependent variable. Hence any decrease (increase) of the error can be interpreted as an increase (decrease) of forecast accuracy. In section 7.3 I make predictions of forecast errors on the basis of the explanatory model. Hence, in that section, the sign of the error was modeled separately.
It should be noted that Smith and Sincich (1988), in an empirical study of the stability over time in the distribution of population errors for states in the USA during the 20th century, find that signed errors are much less stable (and hence much less suitable for predictions) than absolute errors.

For mortality, separate models have been estimated for males and females. The descriptive analysis of mortality errors in section 5.4.2 revealed that female mortality in the Netherlands after 1950 was more difficult to foresee than male mortality. Together with the fact that possible distortions due to (errors in) age structures should be avoided, the life expectancy at birth for males and females was selected for the explanatory model. For males as well as for females, separate models were estimated for the absolute value of the error in the life expectancy and for its sign. The choice for life expectancy as the dependent variable in the context of mortality errors is not without problems. A consequence is that the number of observations is greatly reduced, for two reasons. Firstly, information on the assumed life expectancy at birth is missing for the 1956 forecast. Secondly, as forecast variants for the forecasts of 1972, 1975 and 1980 were formulated in terms of fertility only, mortality errors for the high variant of the 1972 forecast defined in terms of life expectancy coincide with those for the low variant of that forecast. In the case of fertility I have two sets of observations for the forecasts of 1972 and 1975 (high variant and low variant), and three sets for the 1980 forecast (high, medium and low). For errors in the life expectancy, however, there is only one set of observations for each of these three forecasts. Altogether, the data set for mortality consists of 141 observations for males and the same number for females, while that for fertility contains 203 observations. Therefore, the number of independent variables in the models for male and female mortality errors will be kept to a minimum. Table 7.1 presents observed and forecast values of the life expectancy at birth for males and females. Mortality assumptions in the forecasts of 1967 and 1972 were analogous to those in the forecasts of 1965 and 1970, respectively.

Table 7.1.a. Life expectancy at birth, observed and forecasts since 1950, the Netherlands (males)

	Observed	Forecasts 1950	1951	1965 and 1967	1970 and 1972	1975	1980
				years			
1950	70.6	69.4					
1951	70.6	69.4	70.5				
1952	70.6	69.4	70.5				
1953	71.0	69.4	70.5				
1954	71.0	69.4	70.5				
1955	71.0	69.4	70.5				
1956	70.9	69.4	71.3				
1957	71.4	69.4	71.3				
1958	71.5	69.4	71.3				
1959	71.3	69.4	71.3				
1960	71.5	69.4	71.3				
1961	71.5	69.4	72.1				
1962	71.1	69.4	72.1				
1963	71.1	69.4	72.1				
1964	71.2	69.4	72.1				
1965	71.1	69.4	72.1	71.4			
1966	71.1	69.4	72.1	71.4			
1967	71.0	69.4	72.1	71.4			
1968	71.0	69.4	72.7	71.4			
1969	70.9	69.4	72.7	71.4			
1970	70.7	69.4	72.7	71.6	70.6		
1971	71.0	69.4	73.4	71.6	70.6		
1972	70.8	69.4	73.4	71.6	70.6		
1973	71.3	69.4	73.4	71.6	70.6		
1974	71.6	69.4	73.4	71.6	70.6		
1975	71.4	69.4	73.4	71.9	70.2	71.3	
1976	71.5	69.4	74.0	71.9	70.2	71.3	
1977	72.0	69.4	74.0	71.9	70.2	71.3	
1978	71.9	69.4	74.0	71.9	70.2	71.3	
1979	72.4	69.4	74.0	71.9	70.2	71.3	
1980	72.5		74.0	72.3	69.5	71.4	72.4
1981	72.7			72.3	69.5	71.4	72.5
1982	72.7			72.3	69.5	71.4	72.6
1983	72.9			72.3	69.5	71.4	72.7
1984	73.0			72.3	69.5	71.4	72.9
1985	73.1				69.5	71.4	73.0
1986	73.1				69.5	71.4	73.1

Table 7.1.b. Life expectancy at birth, observed and forecasts since 1950, the Netherlands (females)

	Observed	Forecasts 1950	1951	1965 and 1967	1970 and 1972	1975	1980
				years			
1950	72.9	71.5					
1951	72.9	71.5	72.1				
1952	72.9	71.5	72.1				
1953	73.9	71.5	72.1				
1954	73.9	71.5	72.1				
1955	73.9	71.5	72.1				
1956	74.1	71.5	72.8				
1957	74.6	71.5	72.8				
1958	74.9	71.5	72.8				
1959	75.2	71.5	72.8				
1960	75.3	71.5	72.8				
1961	75.8	71.5	73.5				
1962	75.7	71.5	73.5				
1963	75.8	71.5	73.5				
1964	76.2	71.5	73.5				
1965	76.1	71.5	73.5	76.8			
1966	76.1	71.5	74.1	76.8			
1967	76.5	71.5	74.1	76.8			
1968	76.4	71.5	74.1	76.8			
1969	76.3	71.5	74.1	76.8			
1970	76.5	71.5	74.1	77.8	76.3		
1971	76.7	71.5	74.7	77.8	76.3		
1972	76.8	71.5	74.7	77.8	76.3		
1973	77.3	71.5	74.7	77.8	76.3		
1974	77.6	71.5	74.7	77.8	76.3		
1975	77.6	71.5	74.7	78.6	76.5	77.3	
1976	78.0	71.5	75.3	78.6	76.5	77.3	
1977	78.4	71.5	75.3	78.6	76.5	77.3	
1978	78.5	71.5	75.3	78.6	76.5	77.3	
1979	78.9	71.5	75.3	78.6	76.5	77.3	
1980	79.2		75.3	79.4	76.6	77.6	79.1
1981	79.3			79.4	76.6	77.6	79.2
1982	79.4			79.4	76.6	77.6	79.3
1983	79.6			79.4	76.6	77.6	79.4
1984	79.7			79.4	76.6	77.6	79.4
1985	79.7				76.6	77.8	79.5
1986	79.6				76.6	77.8	79.5

International migration was split into immigration and emigration. Forecast errors were computed as the difference between forecast values and observed values for the numbers of immigrants and emigrants. Dependent variables are the absolute value of these errors and the corresponding sign. Forecast errors are limited to the forecasts of 1975 and 1980, as international migration was not included in forecasts produced before 1975 (for a discussion on the implications of this fact for the analysis of forecast errors, see section 6.3.3). Table 7.2 contains observed values and forecast values of immigration and emigration for the period 1976-1986 and the forecasts of 1975 and 1986.

Table 7.2. Immigration and emigration, observed and forecasts, the Netherlands

	Immigration			Emigration		
	Observed	Forecasts		Observed	Forecasts	
		1975	1980		1975	1980
			in thousands			
1976	83.0	84.1		61.5	63.4	
1977	83.9	78.3		61.1	62.9	
1978	89.2	78.6		61.1	61.1	
1979	104.6	75.2		59.8	58.6	
1980	112.5	72.9	112.8	59.5	57.0	59.7
1981	80.2	71.8	82.3	63.2	55.9	62.2
1982	70.7	72.2	80.5	67.5	55.1	60.2
1983	66.8	72.8	78.0	60.8	54.0	59.4
1984	66.9	73.6	75.5	58.9	53.4	58.7
1985	79.4	74.4	73.1	55.2	53.0	58.3
1986	87.4		70.8	54.7		57.9

7.1.2. Independent variables

Not all of the factors which were identified in section 3.4 could be taken into account as independent variables in the explanatory model for observed forecast errors. Several reasons can be given.

The first reason is that some of these factors show too little variation, or too smooth a development in the Netherlands over the period 1951-1986. Examples are population size and population growth. In some cases even there is no variation at all: type of data used, combination of forecasts and frequency of measurement. For instance, the NCBS started to include monthly data in its (short-term) forecasts only recently (De Beer, 1987,

1988), and hence the impact of type of data on accuracy cannot be tested with the data that I have. The second reason is that some factors have to be combined into one overall factor, because they are operationalized into discrete variables (dummy variables) with changes taking place at the same time. For instance, a change in organizational context took place around 1975 (see section 4.5). A dummy variable describing this factor could take on the value of one for errors related to the forecasts of 1975 and 1980, and the value of zero for all other errors. However, such a dummy variable would absorb all other factors which are related to the forecasts of 1975 and 1980, for instance the use of sociodemographic expertise since 1975.
The final reason to exclude certain factors from the model is the lack of data: for international migration forecasts made in 1975 and 1980, the number of data points used in the analysis of trends observed prior to the jump-off year is unknown.

The following independent variables are used to explain the error E(i,j,k) for a forecast with jump-off year i (i = 1950, 1951, 1956, ..., 1980), observed in calendar year j (j = 1951-1986) for a duration of k years (k = 1, 2, 3, ...).
Change in forecast methodology is described by means of a dummy variable DM with a value of zero for all forecasts made between 1950 and 1967, and a value of one for forecasts made in 1970 or later. It will be used only in fertility models.
Forecast's jump-off year is described by means of a set of nine dummy variables F(i) (i=1950, 1951, 1956, ... 1980), each having a value of one for errors related to the forecast with jump-off year i, and zero otherwise. This variable is equivalent to the forecast effect described in chapter 6.
Length of forecast period or LOFP is introduced in the model which explains E(i,j,k) in two ways. The first one is a linear variable LP defined as j-i (j>i), and the second one is SQRLP, the square root of LP. Both variables express LOFP in years.
The *number of data points* used in the extrapolation of fertility and mortality in each forecast or, equivalently, the length of the period upon which the analysis was based is described by means of the variable LAP, which expresses, for each forecast, the length of the period of analysis in years.
The *slope of the observed demographic variable* in question is used to account for a steep increase or decrease of demographic developments. It may have an impact on accuracy at the time the forecast is produced. A positive slope would yield a systematic overestimation, a negative slope an underestimation. Therefore, in the equations for the sign of the errors a variable SF was introduced expressing this slope as the absolute value of the difference between the observed demographic indicator (birth rate, life expectancy, immigrations, emigrations) in calendar year i (when the forecast was produced) and year i-5. Hence the slope variable applies to the *jump-off years* of the forecasts. In the equations for the absolute value of the errors, the slope was assumed to apply to *calendar years,* and it is denoted

by SY. A second variable DTS was substituted for SY in a number of occasions. DTS is a dummy variable which accounts for the trend shift which ocurred in fertility in the Netherlands between 1964 and 1975. Its value is one for forecasts produced between 1965 and 1972, and zero for other forecasts. DTS is only used in fertility models.

Randomness in the observations was operationalized in two ways. A first instability indicator, to be denoted as IIA, is due to Ascher (1978, 51). It focuses on relative changes in observations and may be interpreted as the mean absolute deviation of changes in single-year observations from a ten-year period's "average" change in observation. For a certain jump-off year i and for observations x(i-11), x(i-10), ..., x(i-1) for the ten years preceding i, this indicator is defined as

$$IIA = \sum_{t=1}^{10} \left| \{x(i-1)-x(i-11)\}/x(i-11) - \{x(i-t-1)-x(i-t-2)\}/x(i-t-2)\} \right| /10.$$

Second, a relatively simple instability indicator IIS was used, defined as the standard deviation of observations for the years i-6 to i.

Finally, *technological change* was operationalized as a variable TC having the value 1 for the 1950 forecast, the value 2 for the 1951 forecast, the value 7 for the 1956 forecast, etc. Thus, for any combination of i, j and k, TC is defined as i-1949.

7.1.3. The models

Because I have a time series of observations for each forecast, a combined cross-section and time series model is a natural framework for the models in this chapter. Different degrees of pooling the cross-section and time series data may be used. That is, slope variables and intercept variables may or may not vary over forecasts and/or time (Judge et al., 1985, chapter 13; Maddala, 1983, chapter 14). However, all standard models are applied in a situation where the number of observations on each cross-section unit is the same. But my data set consists of forecasts with six (for the 1980 forecast) to 30 (for the forecasts of 1950 and 1951) observations. To derive estimation procedures for such an "unbalanced" model (Maddala, 1983, 327) would be beyond the scope of the present project (for discussions on this topic, see Judge et al., 1985, 480). More important, however, is the fact that for a number of forecasts the length of the series is simply too short for properly estimating a time-series model. Therefore, the data are simply pooled into one array and only a single equation is estimated each time. A possible consequence is heteroskedasticity: the residuals of the regression models that I estimate may have variances which are different for different forecasts. Possible consequences are briefly discussed below.

Models for the absolute value of the forecast error are formulated in a linear (additive) form, and they are estimated with OLS. Because the dependent variable takes on non-negative values only, the usual assumption of a normal distribution of residuals does not hold. A logarithmic transformation of the dependent variable may be used to create a new dependent variable defined over the entire interval $(-\infty,+\infty)$. Various experiments with the log of the absolute forecast error were carried out, but in many cases I was confronted with heteroskedasticity. On the other hand, in case one estimates a linear model for the original non-negative dependent variable and non-normal residuals using OLS there are a number of implications (see, for instance, Judge et al., 1985, 824):

- the usual parameter estimator is unbiased and consistent, and it has minimum variance among all *linear* unbiased estimators, but it is *not* efficient or asymptotically efficient;
- the usual variance estimator is unbiased and consistent, but *not* (asymptotically) efficient;
- the F- and t-tests on the parameters are not necessarily valid in finite samples, although they have an asymptotic justification.

In general, plots of residuals did not display strong deviations from a normal pattern. Therefore I decided to accept the consequences of a certain degree of non-normality in the residuals, thereby exercising care with F and t tests for small sample sizes. The issue of heteroskedasticity is taken up more formally in section 7.3.

For the sign of the forecast error I use a probit model, as models of this type are particularly well suited for dichotomous dependent variables. In general, the functional form of a probit model for a dichotomous variable y which can take on values 0 and 1 only, is

$$\Pr(y=1|x) = \int_{-\infty}^{\beta x} \{\exp(-u^2/2)/\sqrt{2\pi}\}du = \Phi(\beta x),$$

with x the set of independent variables, β the set of unknown parameters, and Φ the cumulative normal distribution. The probit model uses the normal distribution. As an alternative, I could have estimated the logit model $\Pr(y=1|x) = \exp(\beta x)/\{1+\exp(\beta x)\}$, which uses a logistic distribution. The standardized logistic distribution, having zero mean and variance equal to one, has slightly heavier tails than the standard normal distribution. The logistic and the normal curve are so similar that in practice they yield estimated probabilities that differ by less than 0.02. For discussions on the choice between a probit model and a logit model, see Amemiya (1985, 269) and Aldrich and Nelson (1987, 34).
The probit models in this chapter are estimated by Maximum Likelihood.

7.2 | Results

This section contains estimation results for the absolute errors in the various components. Models for the corresponding signs will be estimated in section 7.3, when they are first needed for prediction of signed errors. For each component I present only the most satisfactory results: estimates of coefficients, corresponding t-values, the adjusted multiple correlation coefficient, and the F-statistic. To obtain the "optimal" equations, backward regression was used. In each step, the least significant variable (at the 5 per cent level) of the previous step is eliminated (the constant is not included in the selection procedure). Three types of models were estimated for the error amplitude: (i) models with dummy variables for jump-off years, (ii) models without jump-off dummies, but with one-year lagged dependent variables, and (iii) models without jump-off dummies and without lagged dependents. Models of type (i) enable us to draw conclusions regarding the impact of changes in forecast approach over the years. As other explanatory variables are included as well, they form a logical extension of the separation models which were estimated in chapter 6. However, they cannot be used for the prediction of errors in future forecasts, since the coefficient of the jump-off dummy of such forecasts is not known. Therefore, models of types (ii) and (iii) were estimated. The reason that I distinguish between these two types is that, from a causal point of view, a lagged dependent variable does not explain much. I expect that the fit of type (ii) models will be better than that for type (iii) models, but the latter models may be preferred for substantive reasons.

7.2.1. Fertility

A first simple assumption is that the absolute errors in birth rates, male and female life expectancies, and migration, are linearly related to the independent variables defined in section 7.1.2. Table 7.3 contains the simple correlation coefficients between independent variables and dependent variables.

A positive (negative) value for a dummy variable F(i) indicates that absolute errors in birth rates in this forecast are above (below) average. The forecasts of 1965, 1967 and 1970 have relatively large errors, and those of 1975 and 1980 were relatively accurate. The coefficient of -0.22 for the methodology dummy leads us to believe that low forecast errors went together with an improved longitudinal approach. The duration variables LP and SQRLP exhibit strong correlation with AEBR, and the signs are as expected. The slope variable SY shows only weak correlation, as opposed to the dummy for the trend shift DTS. Surprisingly enough, the coefficients for the instability indicators IIA and IIS are negative. This can be explained by the fact that the forecast of 1950, with errors below average, was produced right after the post-war baby boom of 1946-47 which brings along a high value of the instability indicators for that forecast. At the same time, the very inaccurate forecast of 1965 was made in a year when a relatively smooth (but slowly rising) trend in

Table 7.3. Simple correlation coefficients between dependent and independent variables for errors in fertility, mortality and international migration

	Absolute errors in birth rates (AEBR)	Absolute errors in life expectancy		Absolute errors in migrations	
		Males (AELEM)	Females (AELEF)	Immig. (AEI)	Emig. (AEE)
F50	-0.09	0.26	0.71		
F51	-0.18	-0.03	0.08		
F56	-0.08				
F65	0.58	-0.37	-0.39		
F67	0.17	-0.34	-0.36		
F70	0.23	0.29	-0.02		
F72	0.05	0.36	0.03		
F75	-0.30	-0.05	-0.09		
F80	-0.28	-0.21	-0.23		
DM	-0.22				
LP	0.42	0.44	0.62	0.37	0.58
SQRLP	0.44	0.45	0.60	0.34	0.53
SY	0.18	0.00	-0.02	0.21	0.27
DTS	0.59				
LAP	-0.65	-0.25	-0.45		
IIA	-0.33	0.19	0.66	-0.21	0.23
IIS	-0.31	0.26	0.75	-0.21	-0.23
TC	-0.01	-0.11	-0.58	-0.21	-0.23
Lagged dependent	0.98	0.96	0.98	0.07	-0.02

births came to an end, see figure 5.3. Perhaps forecasters are much more careful in times of volatile developments than when the trends are very smooth. The one-year lagged error in birth rates is very strongly correlated with AEBR. To calculate the first value of the lagged dependent variable for each forecast, observed errors for duration zero (not contained in the series for $AEBR_t$) are taken as the value for duration one.

Of the estimation results which were obtained I made the selection as shown below (t-values are shown in parentheses).

* with jump-off dummies

$$AEBR_t = \underset{(0.29)}{0.03D50} + \underset{(0.35)}{0.04D51} + \underset{(0.51)}{0.06D56} + \underset{(4.50)}{0.81D65} + \underset{(2.74)}{0.40D67} + \underset{(2.91)}{0.47D70}$$

$$+ \underset{(1.25)}{0.15D72} - \underset{(-0.13)}{0.01D75} + \underset{(0.87)}{0.10D80} + \underset{(8.50)}{0.18SY} + \underset{(54.18)}{0.91AEBR_{t-1}} \qquad (7.1)$$

$R^2(adj.) = 0.97 \quad F = 707$

The one-year lagged dependent variable appears to be highly significant. Its coefficient has the expected sign, and the value is less than one. Forecast duration turns out to be non-significant, similar to the findings in chapter 6. Dummy variables for forecasts other than those of 1965, 1967 and 1970 are not significant. The difference between the estimated coefficients for D67 and D70 is only 0.07, but between 1965 and 1970 the difference is much larger: 0.34. However, the estimated standard error of the latter difference is 0.27, so that the coefficients of the dummy variables for the forecasts of 1965 and 1970 do not differ significantly. When insignificant dummy variables are removed from model (7.1), the estimates for the remaining coefficients are almost unaltered, see expression (7.1'). Therefore, even when one controls for the autoregressive character of AEBR (through the one year lagged dependent variable) and for irregularities in ex-post observed trends (through SY), the 1970 forecast does not show smaller forecast errors than the forecasts of 1965 and 1967, with their relatively simple methodology to extrapolate fertility.

$$AEBR_t = \underset{(4.95)}{0.71D65} + \underset{(2.65)}{0.33D67} + \underset{(2.85)}{0.39D70} + \underset{(11.35)}{0.19SY} + \underset{(69.88)}{0.92AEBR_{t-1}} \qquad (7.1')$$

$R^2(adj.) = 0.97 \quad F = 1799$

* without jump-off dummies

$$AEBR_t = \underset{(2.08)}{0.18} - \underset{(-3.01)}{3.49IIA} + \underset{(9.16)}{0.20SY} + \underset{(74.79)}{0.95AEBR_{t-1}} \qquad (7.2)$$

$R^2(adj.) = 0.97 \quad F = 2207$

Also after controlling for the influence of other variables, the impact of observed instability on errors in birth rates is negative. Including the other instability indicator IIS instead of IIA produces a result which is almost identical to equation (7.2), although the estimated coefficient of IIS is somewhat less significant (t-value is -2.60).

* without lagged dependent variable

$$AEBR_t = \underset{(-1.65)}{-1.71} + \underset{(18.29)}{1.65SQRLP} - \underset{(-3.49)}{0.81IIS} + \underset{(4.45)}{0.28SY} \qquad (7.3)$$

$R^2(adj.) = 0.94 \quad F = 1085 \quad \rho = 0.95$

When the one-year lagged error in birth rates is not included, the Durbin-Watson statistic (not shown here) indicates strong autocorrelation in the residuals. Therefore, equation (7.3) was estimated with the Cochrane-Orcutt

iterative procedure (see, for instance, Maddala, 1983, 278), and estimated t-values are corrected accordingly. (It should be noted that the Durbin-Watson statistic underestimates positive first-order autocorrelation in the residuals, because the data are pooled. Thus the Cochrane-Orcutt procedure was applied in *all* models which do not include the lagged dependent variable. The consequences of pooling for the results of this procedure may be assumed to be of minor importance.)
The estimated correlation between successive residuals is extremely high, as could be expected from expression (7.2): ρ is no less than 0.95. The time-series character of the observations, which was accounted for by the lagged dependent variable in expressions (7.1) and (7.2), is now reflected in the duration variable SQRLP (and in ρ). Its estimated coefficient tells us that a forecast duration of, say, four years produces a rise in forecast error of 1.65 per thousand, as compared with a duration of one year (all other factors being equal). A further extension of the forecast horizon to, say, nine years implies another rise of 1.65 per thousand.
It should be noted that neither the length of the period of analysis (LAP), nor technological change (TC), changes in forecast methodology (DM) or shifts in observed trends (DTS) contribute significantly to the explanation of observed forecast errors in birth rates.

7.2.2 Mortality
The absolute error in the value of the life expectancy for males will be denoted by AELEM, and similarly AELEF for females. There are 141 observations for AELEM and for AELEF. Note that the difference in correlation between F65 and F67 in table 7.3 is only due to differences in jump-off year, as the two forecasts had the same mortality assumptions. The same can be noted for the forecasts of 1970 and 1972. The forecasts of 1950 and, for males, those of 1970 and 1972, exhibit large errors in life expectancy. The forecasts made in 1965, 1967 and 1980 show smaller errors than on average. A very large discrepancy between males and females is found for the 1950 forecast, which was given constant mortality assumptions (see table 7.1). As mortality improved steadily for females and not for males (with an interruption of the upward trend in the life expectancy during the 1960s), the assumption of constant mortality brought about much larger errors for females than for males. But already in the 1960s, the continuous improvement of female life chances was quite well foreseen. The strongly negative correlation (-0.58) between TC and AELEF is another reflection of the trend in forecast errors of female mortality. Other than for fertility, errors in mortality show a good degree of positive correlation with irregularities in observed life expectancies as expressed by IIA and IIS, in particular for females. Also, forecast duration (LP, SQRLP) and length of the period of analysis (LAP) are related to forecast errors, with signs in agreement with a priori considerations. The highest correlations are found again for the one-year lagged dependent variables.

The following estimation results were obtained for males.

* with jump-off dummies

$$AELEMt = \underset{(3.24)}{0.23D50} + \underset{(2.75)}{0.16D51} + \underset{(0.96)}{0.06D65} + \underset{(0.87)}{0.05D67} + \underset{(4.67)}{0.40D70} +$$

$$\underset{(4.76)}{0.45D72} + \underset{(2.93)}{0.24D75} + \underset{(0.11)}{0.01D80} + \underset{(29.98)}{0.90AELEM_{t-1}} \qquad (7.4)$$

$R^2(adj.) = 0.94 \quad F = 254$

As could be expected, the one-year lagged dependent variable is highly significant. Its coefficient has the expected sign, and the value is less than one. Forecast duration has no significant impact, a fact which is similar to earlier findings.

* without jump-off dummies

$$AELEMt = \underset{(2.70)}{0.10} + \underset{(42.25)}{0.99AELEM_{t-1}} \qquad (7.5)$$

$R^2(adj.) = 0.93 \quad F = 1785$

The one-year lagged dependent variable absorbs all variation in AELEM when jump-off dummies are excluded from the analysis. Other explanatory variables have no significant contribution.

* without lagged dependent variable

$$AELEM_t = \underset{(2.88)}{1.92} + \underset{(11.14)}{0.55SQRLP} - \underset{(-3.15)}{0.27IIS} - \underset{(-6.38)}{0.23LAP} \qquad (7.6)$$

$R^2(adj.) = 0.89 \quad F = 364 \quad \rho = 0.94$

Equation (7.6) was estimated with the iterative procedure of Cochrane and Orcutt. A comparison with the results of expression (7.3) shows that the errors in male life expectancies increase three times as slowly (1.65/0.55) as those in birth rates. Note that approximately the same ratio between duration effects in errors for fertility and mortality was found for the separation model in section 6.3.2. Furthermore, an additional decade in the period of analysis reduces the error in the life expectancy of males with 5*0.23 or more than one year. Similar to the case of fertility, errors in male mortality seem to be smaller the larger fluctuations in observed mortality trends are, as is demonstrated by the negative estimate for the effect of IIS. Including

Ascher's instability indicator IIA in model (7.6) instead of IIS resulted in a t-value of only 0.05.

The estimation results for females are shown below.

* with jump-off dummies

$$\underset{}{AELEF_t} = \underset{(3.97)}{0.48D50} + \underset{(3.26)}{0.25D51} + \underset{(0.13)}{0.01D65} + \underset{(0.40)}{0.03D67} + \underset{(3.48)}{0.30D70} +$$

$$\underset{(3.41)}{0.31D72} + \underset{(2.50)}{0.23D75} + \underset{(0.11)}{0.01D80} + \underset{(36.23)}{0.94AELEF_{t-1}} \qquad (7.7)$$

R^2(adj.) = 0.97 F = 618

* without jump-off dummies

$$AELEF_t = \underset{(1.95)}{0.08} + \underset{(67.53)}{1.02AELEF_{t-1}} \qquad (7.8)$$

R^2(adj.) = 0.97 F = 4560

* without lagged dependent variable

$$AELEF_t = \underset{(0.92)}{0.65} + \underset{(8.01)}{0.44SQRLP} + \underset{(7.66)}{1.27IIS} - \underset{(-3.40)}{0.12LAP} \qquad (7.9)$$

R^2(adj.) = 0.95 F = 969 $\rho = 0.94$

The results for females in expressions (7.7)-(7.9) are very similar to those for males in expressions (7.4)-(7.6). However, three points are worth mentioning.

- The estimated effect of the 1950 forecast for females (0.48) is much higher than that of the corresponding effect for males (0.23). The estimated standard error of the difference between these two effects is 0.14. This finding agrees with the observations made above concerning the discrepancies between male and female mortality in the 1950 forecast.
- The one-year lagged dependent variable for females has an estimated effect of 1.02, see expression (7.8). This implies a slow, but unbounded increase of forecast errors. However, in spite of the fact that the corresponding estimate for males is below one in (7.5), the hypothesis that the effect for males is larger than one cannot be rejected at the 5 per cent level.
- The impact of observed instability of female life expectancy on forecast errors in this variable is positive as one would expect, see the estimate of

1.27 in expression (7.9). For males the impact is negative. Therefore, the model contained in (7.6) was estimated without the instability indicator, with the following result:

$$AELEM_t = 1.18 + 0.47SQRLP - 0.16LAP \qquad (7.10)$$
$$(1.91)\ (10.62) \qquad (-5.47)$$

$R^2(adj.) = 0.88 \quad F = 509 \quad \rho = 0.94$

7.2.3. Migration

As explained earlier, only for the forecasts of 1975 and 1980 there are observations on immigration and emigration errors. No migration assumptions were explicitly formulated for other forecasts. This means that there are only 15 observations for forecast durations of at least one year, see table 7.2. Therefore, one should be careful when examining the correlations in table 7.3. There are only two coefficients which exceed 0.5, i.e. those for the correlation between emigration errors and duration (both in linear and square-root form). Hence it is not completely unexpected that the only model which could reasonably be fitted is one for emigration errors and duration:

$$AEE_t = -8.30 + 2.75LP \qquad (7.11)$$
$$(-1.22)\ (2.56)$$

$R^2(adj.) = 0.28 \quad F = 6.56$

This result is quite poor, especially because the estimates in (7.11) are heavily influenced by an outlier for the year 1986 and the forecast of 1975: assumptions regarding international migration in this forecast were only formulated up to 1985, see table 7.2.

For immigration no reasonable model could be fitted. Observed immigration in the Netherlands tends to display a cyclical pattern with a period of approximately six years, superimposed on a long term steady trend (De Beer, 1988, 4; see also section 3.4). As NCBS-extrapolations for immigration (so far) did not take this cyclical pattern into account, forecast errors for immigration show strong irregularities, and with observations for 15 years only one cannot expect to be able to model a time series with a period of approximately six years - that would require a much longer series of observed errors.

7.3 | Predictions

In this section I present out-of-sample predictions for the errors in fertility and mortality. The performance of these predictions will be compared with

the errors observed for the 1984 forecast during the years 1985-1989, which have not been analysed hitherto. This gives an additional check for the robustness of the models in the previous sections.
Some models will have to be re-estimated in slightly different form, so as to make them appropriate for generating predictions. One requirement is that the predicted values of the error amplitudes are non-negative. Another is that the independent variables are such that they can be set exogenously, even for errors which apply to future calendar years. Thus, models which contain the slope variable SY cannot be used for prediction purposes. Finally, residuals should have constant variance, i.e. heteroskedasticity should not occur. (This issue was not considered in the analyses of section 7.2 in order to keep the number of necessary computations limited.) The likelihood-ratio test will be used to check for heteroskedasticity. To that end, residuals (from OLS regression) belonging to the same forecast are grouped together and the residual variances are estimated for each forecast. Next these estimated variances are compared with the estimated variance obtained from the entire sample (see, for instance, Maddala, 1983, 263). The likelihood-ratio statistic has a χ^2-distribution with g-1 degrees of freedom, where g is the number of groups. When the hypothesis of homoskedasticity has to be rejected, parameter estimates can be obtained by means of weighted least squares, with residual variances from OLS in each group as inverse weights.
For fertility as well as for mortality I will select at least two models: not only one or more *with* the dominant lagged dependent variable, but also at least one model *without* that variable. Models of the latter type are used for the predictions because, from a causal point of view, they provide more insight than those in which the lagged dependent variable dominates. A comparison of the predictions from more than one model for the same component gives an indication of the confidence one can have in those predictions.
Given the poor results for immigration and, to a certain extent, also for emigration, no predictions will be generated for errors in these components.

7.3.1. Fertility

Equation (7.2) was re-estimated without the slope variable SY. The result is

$$AEBR_t = \underset{(4.75)}{0.47} - \underset{(-1.96)}{2.69IIA} + \underset{(63.90)}{0.96AEBR_{t-1}} \qquad (7.12)$$

R^2(adj.) = 0.96 F = 2311

The instability indicator IIA has, again, a negative impact on fertility errors. Although no negative predictions resulted for the 1984 forecast, there is no guarantee that this will also be the case for other forecasts. Therefore the reciprocal value of IIA was taken as the independent variable. This resulted in

$$AEBR_t = 0.18 + 0.01(IIA)^{-1} + 0.94AEBR_{t-1} \quad (7.13)$$
$$\quad\quad (2.41) \quad (3.77) \quad\quad\quad (55.47)$$

$R^2(adj.) = 0.96 \quad F = 2433$

Model (7.3) was re-estimated (using the Cochrane-Orcutt procedure) without the negative constant and SY, and with the reciprocal of IIS, see expression (7.14).

$$AEBR_t = 0.15(IIS)^{-1} + 1.40SQRLP \quad (7.14)$$
$$\quad\quad (2.33) \quad\quad\quad (18.54)$$

$R^2(adj.) = 0.94 \quad F = 2969 \quad \rho = 0.97$

Models (7.12)-(7.14) were checked for heteroskedasticity. The highest value of the likelihood-ratio statistic is 18.5, for model (7.14). This is below the critical value of 21.0, which is the 95-th percentile of the χ^2-distribution with 12 degrees of freedom. Thus, the assumption of equal residual variances for the 13 forecasts cannot be rejected.

The sign of the error in the birth rate (SEBR) was explained with the probit model described in section 7.1.3. In case of an overestimation, SEBR was given the value of one, and zero otherwise. Possible explanatory variables are the sign one year earlier ($SEBR_t$-1) and the slope of the observed birth rate during the five years preceding the jump-off date (SF). The latter variable turns out to be not significant, and the following ML-estimates are obtained (estimated standard errors are given between parentheses):

$$Pr(SEBR_t=1|SEBR_{t-1}) = \Phi(-0.96 + 2.98SEBR_{t-1}). \quad (7.15)$$
$$\quad\quad\quad\quad (0.18) \quad (0.30)$$

Hence, subsequent errors in birth rates are not only highly correlated in absolute value, as shown in expressions (7.1) and (7.2), but also in terms of their sign. When an overestimation is recorded for a certain year, the probability that there will be an overestimation again in the following year is nearly 98 per cent, as $\Phi(-0.96+2.98)=0.978$.
The simplicity of the type of model given by expression (7.15) facilitates, along a different line of thought, a prediction for the probability for an overestimation, given the sign of the forecast error in the preceding year. Because only the sign in the previous year turns out to be relevant for the explanation of the sign in the current year, one could record for the error in

birth rate *including its sign* how often a positive error (overestimation) was followed by a negative one (underestimation), and vice versa. Table 7.4.a shows that this occurred in 3 cases in the data set. The relative frequency of an overestimation in two subsequent years is, therefore, 135/138 = 97.8 per cent, which agrees with the results of the probit model. In other words, the Maximum Likelihood estimates of -0.96 and 2.98 coincide with those found with the Method of Moments, i.e. directly from corresponding observations. More generally, let f_{++} denote the relative frequency of an overestimation in two subsequent years, as a proportion of all cases with an overestimation in the preceding year, and similarly let f_{-+} denote the relative frequency of an underestimation followed by an overestimation, as a proportion of all cases with an underestimation in the first year. Thus, in table 7.4.a, f_{++} = 135/138, and f_{-+} = 11/65. Then the parameters of the model $Pr(SEBR_t=1|SEBR_{t-1}) = \Phi(\beta_0+\beta_1 SEBR_{t-1})$ can be estimated with the method of moments as

$$b_0 = \Phi^{-1}(f_{-+}),\ b_1 = \Phi^{-1}(f_{++}) - b_0. \qquad (7.16)$$

The fact that the estimators in (7.16) are ML-estimators can be demonstrated as follows. Consider for the i-th observation the model

$$Pr(y_i=1|x_i) = p(x_i,\beta),$$

where p is an unspecified function between zero and one. Then the normal equations are (Judge et al., 1985, 765)

$$\frac{d\log L}{d\beta} = \sum_{i=1}^{N} \frac{y_i}{p(x_i,\beta)} \frac{dp(x_i,\beta)}{d\beta} - \frac{1-y_i}{1-p(x_i,\beta)} \frac{dp(x_i,\beta)}{d\beta} = 0.$$

Now x_i is a vector of two dummy variables, one for the constant term β_0 (with only one's), the other for the coefficient β_1. Denote the probability $p(x_i,\beta)$ for x_i=1 as p_1. Then, as Evert van Imhoff pointed out to me, the normal equations reduce to a very simple form. For β_1 one can write

$$\frac{d\log L}{d\beta_1} = \sum_{i=1}^{N} \frac{y_i}{p_1} \frac{dp_1}{d\beta_1} - \frac{1-y_i}{1-p_1} \frac{dp1}{d\beta_1} = 0, \text{ or}$$

$$\frac{dp_1}{d\beta_1} \left[\sum_{i=1}^{N} \frac{y_i}{p_1} - \frac{1-y_i}{1-p_1}\right] = 0.$$

Thus, provided that $dp_1/dß_1 \neq 0$, this implies that the ML-estimator for p_1 equals f_{++}. Similar reasoning holds for the constant $ß_0$.
Finally note that the form of $p(x_i,ß)$ was left unspecified here. Therefore, moment estimators for the relevant probabilities also coincide with ML-estimators in other models, for instance in the logit model (provided that there are not more than two regressors, both in dummy form, and that the inverse function p^{-1} exists).

Table 7.4.a. Classification of overestimations and underestimations of birth rates, forecasts since 1950

year t	year t-1 overestimation	 underestimation	total
overestimation	135	11	146
underestimation	3	54	57
total	138	65	203

Table 7.4.b. Classification of overestimations and underestimations of birth rates, forecasts 1975 and 1980

year t	year t-1 overestimation	 underestimation	total
overestimation	26	2	28
underestimation	2	10	12
total	28	12	40

Table 7.5 shows predicted errors for the 1984 forecast, based on predictions according to models (7.12)-(7.14). Predictions were generated recursively (except for equation (7.14)) for the years 1985-2000, and to start up the process, observations in 1984 were used for the lagged values of the sign of the error and its absolute value. As the observed birth rate in 1984 was 0.03 per thousand lower than the value for that year which was foreseen in the medium variant of the 1984 forecast, the probability of an underestimation in the years from 1985 onwards is – see expression 7.15 – $1-\Phi^{-1}(-0.96)=0.83$ (or 54/65, see table 7.4). Therefore, there is an 83 per cent chance that the predicted absolute errors in table 7.5 represent underestimations.

The predictions are quite similar: they show a steady growth of the absolute error, from a level between 0.5 and 0.6 to an extremely high level of 4.4-5.2 per thousand. For the first three years, the observed errors agree quite well with the errors which were observed for the medium variant of the 1984 forecast. However, contrary to the predicted errors, the rise in the observed

Table 7.5. Predicted absolute values of errors in birth rates according to some models, and observed for forecast 1984 (medium variant)

	Model (7.12)	Model (7.13)	Model (7.14)	Observed, forecast 1984	Model (7.17)	Model (7.18)
			per thousand			
1984				0.03		
1985	0.45	0.53	0.62	0.36	0.16	0.23
1986	0.85	1.00	1.06	0.85	0.31	0.47
1987	1.23	1.45	1.44	0.96	0.45	0.66
1988	1.60	1.86	1.78	0.92	0.58	0.83
1989	1.96	2.24	2.08	1.06	0.68	0.97
1990	2.30	2.60	2.35		0.76	1.09
1991	2.63	2.94	2.61		0.82	1.21
1992	2.95	3.26	2.85		0.87	1.31
1993	3.26	3.55	3.08		0.91	1.41
1994	3.55	3.83	3.30		0.94	1.50
1995	3.83	4.09	3.51		0.96	1.58
1996	4.10	4.33	3.71		0.97	1.66
1997	4.37	4.55	3.90		0.99	1.73
1998	4.62	4.77	4.09		1.00	1.81
1999	4.86	4.96	4.27		1.00	1.87
2000	5.09	5.15	4.44		1.01	1.94

errors levels off at values between 0.92 and 1.06 towards the end of the 1980s. The reason that models (7.12)-(7.14) produce errors with too large amplitudes is that they were fitted for a historical period which saw relatively large errors. Therefore, I re-estimated these three models for the "smooth" period 1976-1986, with observations for the forecasts of 1975 and 1980. Results are shown below.

$$AEBR_t = 13.09IIA + 0.75AEBR_{t-1} \tag{7.17}$$
$$\qquad (1.98) \qquad (5.34)$$

R^2(adj.) = 0.83 F = 186 $\rho = 0.54$

$$AEBR_t = -1.14 + 0.45VAR + 0.53SQRLP \tag{7.18}$$
$$\qquad (-2.87) \quad (4.27) \qquad (4.82)$$

R^2(adj.) = 0.79 F = 72 $\rho = 0.76$

Note the differences compared with the results for the longer period. Firstly, the time-series character is much weaker, see the relatively low values of ρ, of the coefficient of the lagged dependent variable and of that coefficient`s t-value. Secondly, the impact of duration on errors in birth rates, as exhibited by the coefficient of SQRLP, is much smaller. These two differences can be explained by the dominant character of the forecasts of 1965 and 1967 with large and rapidly increasing overestimations of the birth rate. A third difference is that the coefficient of the instability indicator has the expected sign here, while it is negative for long-term errors. The reason is that the forecasts of 1950 and 1965, which are to a large extent responsible for the inverse relationship between instability and error size (see the discussion of table 7.3 in section 7.2.1), are no longer contained in the short-term analysis. A final difference between short-term results and those for the long term is that a significant influence is exercised by a variable which accounts for the type of fertility variant: high, medium, or low. Because (fertility) variants were introduced only in 1972 in Dutch demographic forecasts, such a variable was not necessary to explain the long-term errors. VAR is a variable which has simply the values of 3, 2 and 1 for forecast errors related to high, medium and low variants respectively. (Variables HVAR and LVAR which were used in chapter 6 to account for the influence of forecast variants on forecast errors turned out to be non-significant.)

Predictions according to (7.17) en (7.18) are given in the last two columns of table 7.5. Autocorrelation in the residuals was taken into account for these predictions (see Judge et al., 1985, 316). The predicted errors in birth rates rise much slower now than for models (7.16)-(7.18). The autoregressive term $0.75AEBR_{t-1}$ in (7.17) causes AEBR to level off, and simple calculation shows that the equilibrium value equals 1.024. Predictions according to model (7.18) are very close to observations, while those generated by model (7.17) tend to be a bit low. Table 7.4.b shows that for short-term errors, the probability of an underestimation in a certain year following an underestimation in the previous year can be estimated as 10/12 or 83 per cent. Note that this is the same value as that found for the long term.

7.3.2 Mortality

Two models were used to generate predictions for the absolute error in the life expectancy of males. The first one is model (7.5), while the second one was derived from expression (7.10). The latter model was re-estimated with 1/LAP as an independent variable, because periods of analysis larger than some 15 years will create negative predicted values of AELEM for short durations. The result is

$$AELEM_t = \underset{(-2.42)}{-1.07} + \underset{(9.50)}{0.47SQRLP} + \underset{(3.47)}{5.35(LAP)^{-1}} \qquad (7.19)$$

$R^2(adj.) = 0.86 \quad F = 435 \quad \rho = 0.92$

For females, I selected model (7.8) as the first model. Model (7.9) was re-estimated with 1/LAP, for reasons just described. This resulted in

$$AELEF_t = \underset{(-2.41)}{-1.32} + \underset{(8.14)}{0.45}SQRLP + \underset{(4.24)}{0.99}IIS + \underset{(2.97)}{7.43}(LAP)^{-1} \qquad (7.20)$$

$R^2(adj.) = 0.95 \quad F = 938 \quad \rho = 0.94$

Models (7.5) and (7.19) for the life expectancy of males, and models (7.8) and (7.20) for that of females were checked for heteroskedasticity. The highest value of the likelihood-ratio statistic was 11.4, which is below the critical value of 14.1 for the χ^2-distribution with 7 degrees of freedom. Thus the hypothesis of equal variances over the forecasts cannot be rejected.

The sign of the error in the life expectancy at birth (SELEM for males and SELEF for females) was linked to independent variables through a probit model. For males as well as for females, the slope of observed life expectancies in the jump-off year of the particular forecast appeared to be non-significant. Thus the only "independent" variable included in the probit model is the one-year lagged sign. Results for males and females are shown below, with estimated standard errors in parentheses.

$$Pr(SELEM_t=1|SELEM_{t-1}) = \Phi(\underset{(0.21)}{-1.66} + \underset{(0.32)}{2.71}SELEM_{t-1}) \qquad (7.21)$$

$$Pr(SELEF_t=1|SELEF_{t-1}) = \Phi(\underset{(0.25)}{-1.92} + \underset{(0.37)}{3.11}SELEF_{t-1}) \qquad (7.22)$$

The estimated probability of an overestimation in two subsequent years is 85.3 per cent (=$\Phi(1.05)$) for males and 88.3 per cent (=$\Phi(1.19)$) for females. On the other hand, the probability of an underestimation in a certain year followed by an overestimation in the next year is only 4.9 per cent for males and 2.7 per cent for females, see also table 7.6.

The results in table 7.7 show that predictions for errors of the female life expectancy agree quite well with observed values in the 1984 forecast. The fit is somewhat better for model (7.20) without the lagged dependent variable, than for model (7.8) with such a variable. For males the results are disappointing. The predictions imply a monotone rise from 0.3 years in 1985 to a level between 1.2 and 1.7 years in 2000. On the other hand, observed errors in male life expectancy fluctuate somewhat during the years 1984-1989, around a level which is much lower than that of the predictions. To test whether these large predicted errors are caused by the relatively large

Table 7.6. Classification of overestimations and underestimations of life expectancies

	year t-1		
	overestimation	underestimation	total
	males		
year t			
overestimation	35	5	40
underestimation	6	98	104
total	41	103	144
	females		
year t			
overestimation	30	3	33
underestimation	4	107	111
total	34	110	144

Table 7.7. Predicted absolute values of errors in male and female life expectancies according to some models, and observed for forecast 1984 (medium variant)

	Males			Females		
	Model (7.5)	Model (7.19)	Observed, forecast 1984	Model (7.8)	Model (7.20)	Observed, forecast 1984
1984	0.20	0.03				
1985	0.30	0.32	0.20	0.18	0.41	0.38
1986	0.40	0.41	0.36	0.26	0.50	0.34
1987	0.50	0.50	0.12	0.34	0.58	0.44
1988	0.59	0.57	0.11	0.42	0.65	0.49
1989	0.68	0.64	0.12	0.50	0.71	0.73
1990	0.78	0.70	0.59	0.77		
1991	0.87	0.76	0.67	0.83		
1992	0.96	0.82	0.76	0.88		
1993	1.05	0.87	0.85	0.93		
1994	1.14	0.93	0.94	0.98		
1995	1.23	0.98	1.03	1.02		
1996	1.32	1.03	1.13	1.07		
1997	1.41	1.08	1.22	1.11		
1998	1.49	1.13	1.32	1.16		
1999	1.58	1.17	1.42	1.20		
2000	1.66	1.22	1.52	1.24		

forecast errors in male life expectancy in the forecasts of 1965, 1967, 1970 and 1972 during the late 1960s and the 1970s, I re-estimated models (7.5) and (7.19) for the errors of the forecasts of 1975 and 1980. Predictions which were subsequently made with these short-term models are close to those produced by models (7.5) and (7.19), so that an improvement of the out of sample predictions for errors in male life expectancies could not be obtained. On the other hand, this shows that models (7.5) and (7.19) are quite robust, and it suggests that the low errors in male life expectancy connected with the 1984 forecast as observed for the years 1987-1989 are a-typical. Indeed the observed male life expectancies for these three years (73.5, 73.7 and 73.9 years, respectively) are about 0.3 to 0.4 years higher than one would expect from a simple extrapolation of the values since the 1970s.

7.4 | Conclusions

The results of this chapter may be summarized as follows. Forecast errors in birth rates and life expectancies at birth show a great deal of inertia. The absolute value of the error in a certain year is, to a great extent, determined by the absolute error in the preceding year. At least 92 per cent of the variation in the long-term error for birth rates and life expectancies could be "explained" this way, and 83 per cent of the short-term error. As an alternative, the length of the forecast period can serve as a good explanatory variable for the errors in fertility forecasts and mortality forecasts. A linear form and a square-root form of this variable work approximately equally well. Errors in birth rates grow about three times as fast when forecast duration increases as errors in life expectancy. For the more recent forecasts (those produced since 1975), length of forecast period turned out to be non-significant for fertility errors, while it is still a good explanatory variable for errors in life expectancy. The number of data points contained in the analysis before a forecast is produced, or, in other words, the length of the period upon which the extrapolation of demographic indicators is based contributes also to the explanation of mortality errors: the longer this period, the smaller the errors in life expectancies. However, no effect could be established upon the accuracy of fertility forecasts.

Differences between fertility and mortality in the explanation of ex-post observed errors are striking. The slope (positive or negative) of observed trends influences fertility errors, but not those for mortality. A possible explanation is that, on the whole, observed life expectancy at birth in the Netherlands is steadily increasing, with only minor fluctuations around a stable trend (except for a temporary decline in male life expectancy around 1970). On the other hand, birth rates declined dramatically between 1964 and 1975, and they have more or less stabilized since the mid-1970s. The impact which a possible instability of a demographic indicator during the years prior to the first forecast year may have on that forecast's performance led to puzzling findings. The impact is positive for errors in female life expectancies, and for short-term errors in birth rates, as could be expected on a priori grounds. However, for the life expectancy of males, and for fertility on the long run, forecast errors tend to be reduced when observed indicators become more unstable. It seems as if forecasters are sometimes more cautious than in other instances.

No significant impact could be traced of the change in methodology which took place between 1967 and 1970, when a cohort approach for the extrapolation of fertility was introduced, based on marriage duration. When we control for disturbing effects of strongly decreasing ex-post observations, and when the time-series character of errors in birth rates is taken into account, the errors in birth rates for the 1970 forecast are lower than those for the 1965 forecast, but the difference is not significant. The errors for the 1970 forecast are even higher than those of 1967.

Errors in numbers of immigrants could not be modeled, and those for emigrants could only to a very limited extent be explained by forecast duration. Other factors turned out to be of no significance for emigration errors.

8. CONFIDENCE INTERVALS FOR FORECAST RESULTS

Confidence intervals around forecast results give an appropriate refelection of the ex-ante uncertainty which the forecaster expects to be connected to the forecast.

> " ... Even though we may never be able to forecast future *populations* with a high degree of accuracy, we may be able to develop relatively accurate forecasts of the *distribution of errors* surrounding our point forecasts. Indicating the potential range of errors surrounding population forecasts may be the most useful service the producers of population ... (forecasts) ... can provide to their users..." (Smith and Sincich, 1988, 473).

A confidence interval consists of three elements: an upper boundary, a lower boundary, and a probability that the actual figure will lie between these two boundaries. Thus, the concept of a confidence interval is a logical extension of the notion of forecast variants, in the sense that it attaches a certain probability to the interval between the high variant and the low variant (cf. chapter 2, sections 2.3.1 and 2.3.3).

In this chapter I apply two methods to construct confidence intervals around forecast results: (i) the method of Stoto (1983), which is used for total population size; and (ii) the method of Cohen (1986), which is appropriate for age structures and total population size.

Section 8.1 contains a brief introduction to the two methods to construct confidence intervals. An application of these methods to total population size forecasts for the Netherlands is given in section 8.2. More detailed results on intervals around forecasts of a few selected age classes are presented in section 8.3.

8.1 | Introduction

Several methods exist to generate confidence intervals around results of population forecasts. Cohen (1986, 105) classifies these methods in two groups: empirical methods, based on the distribution of errors of old forecasts; and model-based methods, derived from an explicit stochastic model of population development. The method of Stoto is an empirical method, while that of Cohen is model-based. The two approaches are particularly suited for application, compared to other methods, because of their computational simplicity (both methods) and (Stoto's method) because the error-distributional aspects have been studied intensively for a variety of situations.

* Stoto's method
  Stoto's method rests on the assumption that the future distribution of the error in the annual growth rate of total population is the same as the current distribution of this error. He established empirically the fact that these errors are approximately normally distributed. Given a forecast, he constructs an approximate two-thirds confidence interval for the true growth rate as the average projected growth rate plus or minus one standard deviation of the error in the projected growth rate.
  Stoto (1983) suggests for developed countries taking an estimated standard deviation of the error (SDE) of 0.3 per cent. However, the analysis in chapter 5 of ex-post observed errors in annual population growth revealed that this value would be too high for the Netherlands, see table 5.7. I assume that SDE = 0.1 per cent is a more appropriate value for that country. I will apply an SDE of 0.1 per cent to the medium variant of the NCBS-forecast produced in 1989.

* Cohen's method
  The method described by Cohen (1986) rests on the assumption that the sequence of transition probability matrices **P**(1), **P**(2), **P**(3), ... in the familiar matrix model for the projection of an age-structured population **K**(t+1)=**P**(t+1).**K**(t) is, what Cohen calls, a stochastically stationary ergodic sequence. (Net migration between t and t+1 is assumed to be a linear function of **K**(t), and hence it can be absorbed into the elements of **P**(t+1).) Stochastically stationary means that the joint probability distribution of any finite number of the matrix **P** is invariant with respect to shifts in time. (By contrast, demographically stationary refers to a population which is constant exactly or on average in total size and in age-groups.) The sequence {**P**(t)} is ergodic if, for every event defined in terms of a finite number of matrices **P**, the average frequency of the event converges almost surely to the probability of the event. A detailed discussion of further technicalities involved in Cohen's model would be beyond the scope of this chapter. The interested reader is referred to Cohen's paper, and the references therein.
  The total population at time t is written as Z(t)=**1'K**(t), where **1** is the vector with every element equal to one. Cohen (1986, 107) states that the method to produce point forecasts and corresponding confidence intervals does not apply only to total population size, but to any linear function **a'K**(t) of the population vector **K**(t), where **a** is a vector with non-negative elements, and at least one element strictly positive. In my application

of Cohen's method to Dutch data, I shall analyse total population size and a few selected age-groups.
The natural logarithm of total population size at time t is represented as W(t). Then, with probability one, W(t) asymptotically (that is, after a long time) grows linearly at a fixed rate per unit time. Denote this rate by log(λ). For the stochastic model described by Cohen, log(λ) is the precise analog of the intrinsic rate of natural increase r in the classical stable population model. Thus, if log(λ) is positive, then on average the population is growing. If log(λ) is negative, the population is decreasing. And for log(λ)=0, the total population size is constant on average.
The quantity W(t)-t.log(λ) is asymptotically normally distributed with mean 0 and standard deviation equal to $\sigma\sqrt{t}$. Cohen notes that stable population theory may be considered a special case of this model in which σ=0.
For a population forecast with jump-off year t=T, and base year (first year of observation) t=1, Cohen proposes the following estimators log(L) and s for log(λ) and σ, respectively

$$\log(L) = [W(T) - W(1)]/(T - 1), \tag{8.1}$$

$$s = 1/2\sqrt{(1/2\pi)}.\left\{\frac{\sum_{j=1}^{T-1} j^{-3/2} \,|\, W(1+j) - W(1) - j.\log(L)|}{\log(T-1)} + \right.$$

$$\left. + \frac{\sum_{j=1}^{T-2} j^{-3/2} \,|\, W(2+j) - W(1) - j.\log(L)|}{\log(T-2)}\right\}, \; T \geq 5. \tag{8.2}$$

The estimator log(L) for log(λ) is a maximum likelihood estimator. For s we only know that it is a consistent estimator (Cohen, 1986, 108). For values of T<5, the estimator s makes no sense.
The estimator s is based on weighted absolute deviations, and weights with exponent -3/2. Cohen refers to a study by him and Heyde in which it is shown that the more usual estimator of a standard deviation, i.e. one which is based on the square root of the weighted squared deviations, and weights with exponent -1, is not consistent for σ. For some future time τ>T, the maximum likelihood estimator w(τ) for W(τ) is

$$w(\tau) = W(T) + (\tau - T).\log(L)$$

$$= W(T) + \frac{(\tau - T)}{(T - 1)}.[W(T) - W(1)]. \tag{8.3}$$

An approximate 100(1- α) per cent confidence interval for $w(\tau)$ is given by

$$w(T) + \frac{(\tau - T)}{(T - 1)} [W(T) - W(1)] \pm s.[\frac{(\tau - T)^2}{(T - 1)} + \tau - T]^{1/2} z_{1/2\alpha} . \qquad (8.4)$$

Here, $z_ß$ is a number such that a fraction ß of the probability density of a standard normal distribution is to the right of $z_ß$. Thus, for a ß of $2^{1/2}$ per cent, the corresponding z-value is 1.96.

The confidence interval estimator given in (8.4) is Cohen's estimator no. 2. He discusses an alternative estimator (the Heyde-Cohen estimator) which is somewhat wider, and which is tedious to compute. Cohen interprets estimator (8.4) as giving a minimum level of uncertainty associated with exponential point estimates, although he also suggests that the Heyde-Cohen estimator produces unnecessarily wide intervals in his numerical example of the population of Sweden over the period 1880-2005 (Cohen, 1986, 121, 122).

8.2 | Total population

I applied the method of Stoto to the medium variant of the forecast of 1989 for the Netherlands, which was published in December 1989 by the NCBS (see also NCBS, 1990). From table 5.7 it can be seen that Stoto's suggestion to take a value of 0.3 per cent for the standard deviation of the error in annual population growth (SDE) is too pessimistic for the Netherlands: the average SDE for all forecasts since 1950 is only 0.12 per cent, and for the second half of the 1970s the SDE was at most 0.07 per cent. I assume that an SDE of 0.1 per cent gives realistic estimates (given the model) for the confidence interval around point estimates of current forecasts of the Netherlands. The mean error in annual population growth (ME) is 0.03 per cent for the forecasts 1950-1980, see table 5.7. The ME of the 1975-forecast was negative, that of the 1980-forecast was positive. The average annual population growth in the 1989-forecast (with jump-off date 1 January 1989) starts at a level of 0.29 per cent for the year 1989, then it increases to 0.49 per cent for the period 1989-1999, and next it declines steadily to levels of 0.20 per cent (1989-2030) and 0.05 per cent (1989-2045). Thus, the observed ME of 0.03 per cent is much smaller than the average growth rates of the NCBS-forecast (except perhaps for the years beyond 2030). Therefore a correction of these growth rates with an ME of around 0.03 per cent would not change very much the boundaries of the confidence interval (growth rate plus or minus 0.1 per cent for a probability of 68 per cent). Thus I left the NCBS growth rates unchanged.

In table 8.1 approximate two-thirds (68 per cent) confidence intervals are presented for total population size for the Netherlands in the period 1989-2050. The results indicate a clearly growing interval as we look further into the future: 290 thousand in 1995, 375 thousand in 2000, 540 thousand in 2005, almost 1.2 million in 2025, and a good 1.8 million in 2050. A comparison between these confidence intervals and the intervals between the high and the low variant of the NCBS-forecast reveals that the NCBS-interval is very narrow in 1990, but that it widens very rapidly after that year. This means, given the assumptions of Stoto's model, that the ex-ante probability that the actual population size will fall between the high and the low variant is much lower than two-thirds in 1990, while this probability grows rapidly after that year. For instance, in 1990 it is only 30 per cent, in 1991 it is just below two-thirds (64 per cent), and next it increases to 80 per cent in 1992, and 94 per cent in 1995, to values close to one in the years thereafter. In other words, the high-low margin is much too narrow for the first year, and it is unnecessarily wide for the years after 1995.

Table 8.1. Approximate two-thirds confidence intervals according to Stoto's method, and high and low variants of NCBS-forecast, total population size 1989-2050, the Netherlands

	Stoto's method		NCBS-forecast		
	Upper	Lower	High	Low	Probability
			thousands		
1989	14,805	14,805	14,805	14,805	
1990	14,922	14,862	14,898	14,886	0.30
1995	15,411	15,120	15,478	15,129	0.94
2000	15,884	15,510	16,095	15,296	0.98
2005	16,263	15,722	16,613	15,372	0.99
2010	16,499	15,792	16,983	15,315	0.99
2015	16,645	15,773	17,260	15,172	
2020	16,742	15,707	17,506	14,966	
2025	16,791	15,596	17,744	14,683	
2030	16,767	15,418	17,951	14,303	0.99
2035	16,639	15,148	18,072	13,821	
2040	16,405	14,786	18,083	13,253	
2045	16,114	14,379	18,033	12,647	
2050	15,820	13,976	17,987	12,048	0.999

The extremely wide margins for the period beyond 1995 attached to the NCBS-forecast of 1989 contrast sharply with those of the 1984-forecast, for

which I found earlier a high-low margin which corresponds closely to a two-thirds confidence interval (Keilman, 1986, 70). Although I made a somewhat pessimistic choice of SDE=0.15 per cent on that occasion, this explains only part of the difference between the probabilities of the 1984-forecast (68 per cent) and those of the 1989-forecast (95-99 per cent for forecast periods of seven years or more). Furthermore, investigations for the USA have shown that the high-low margin of USBC-forecasts for a forecast period up to some 15 years corresponds to approximate two-thirds confidence intervals, too (Stoto, 1983, 18; USBC, 1989, 15). The relatively low value of the SDE which is characteristic for Dutch forecasts implies that most of the probability density is concentrated in a narrow band around the medium variant. Except for the first few years, the NCBS may very well reduce the high-low margin by as much as 50 per cent of its present value.
The ex-ante probability that the population size on January 1st 1990 will fall between the margins of the high and the low variant of the 1989-forecast (14.898 million and 14.886 million, respectively) was estimated earlier to be only 30 per cent. The actual population size at that date turned out to be 14.892 million. One may say that the forecasters at the NCBS were just lucky that they hit the target in spite of the relatively small probability of success. On the other hand, the forecast was released in December, and it was prepared during Autumn 1989. At that time, preliminary figures regarding live births, deaths, and migrations were already known for the first eight to nine months of the year, and this information was used in the forecast. This explains why the margin for January 1st 1990 could be relatively narrow, and had to be much wider for 1991 (64 per cent)

I now describe an application of Cohen's method to Dutch data for the period 1830-1988. Data were taken from NCBS (1970) and from various issues of the "Maandstatistiek van de Bevolking". I have two data sets. The first one gives information on total population size for census dates 1830, 1840, 1849, 1859, 1869, 1879, 1889, and 1899; and for calendar years (from civil registration) 1910, 1920, ... , 1980. Censuses in the years from 1849 to 1899 took place in the months of November or December, and on four occasions even on December 31st. Therefore I consider census data as decennial, applying to January 1st. The second data set consists of total male and total female population for each of the years (january 1st) for the period 1900-1989.

A key assumption for this method is that the process which generates the vital rates contained in the sequence $\{\mathbf{P}(t)\}$ is (stochastically) stationary. To test this assumption, I investigated whether the growth rate log(k) is, on average, constant. Following Cohen's suggestion, log(L), the estimator of log(k), was computed according to expression (8.1) for a fixed jump-off year T, and various base years t. With the decennial data, I calculated log(L) for T = 1990 and t = 1830, 1840, ... , 1980. For the annual data I selected the jump-off year T as 1989 and the years 1900, 1901, ... , 1988 as base years. Both on the long term and on the short term there is a clear tendency of a falling growth rate in the Dutch data. The decline starts around 1880, see figure 8.1. Additional evidence of non-stationarity was found when first differences in the logarithm of population size (W(t+1)-W(t)) were calculated. In both data sets, a strong decline in these differences are displayed after the 1950s (data not shown here). Stationarity would imply a constant difference (on average). Thus the conclusion is that the assumptions necessary for applying Cohen's model to total population in the Netherlands do not hold, and that his approach cannot be used.

Figure 8.1.a. Estimated growth rates (log L) of total population size in the Netherlands, jump-off year 1990, base years since 1830

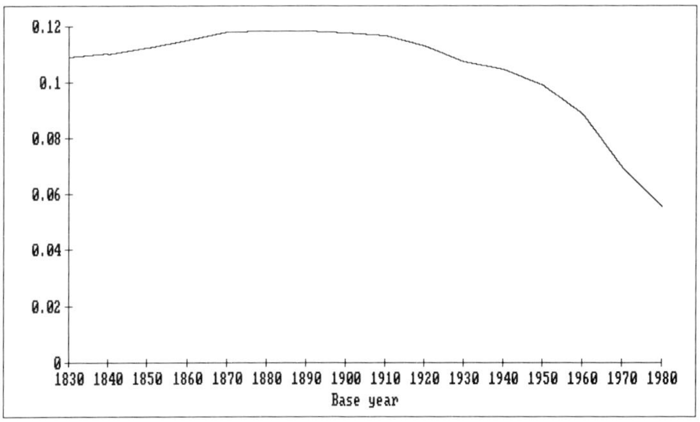

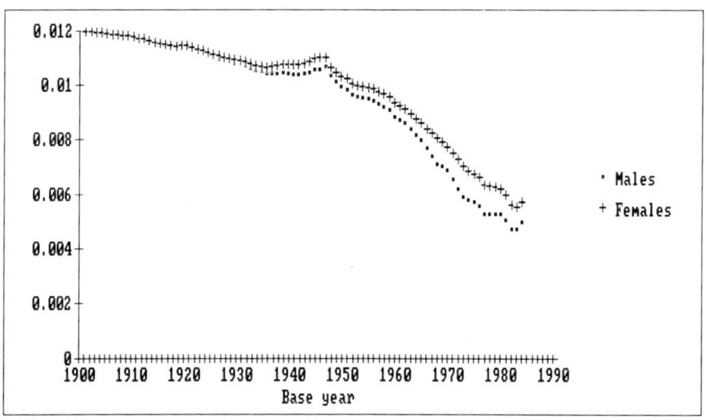

Figure 8.1.b. Estimated growth rates (log L) of total population size in the Netherlands, jump-off year 1989, base years since 1900

Cohen applied his method to the case of Sweden. He had data for total population at five-year intervals for the period 1780-1980. It is remarkable that Cohen accepted the stationarity assumption to hold for the period 1880-1980. In his discussion he states that the "... evidence against stationarity is not conclusive in the absence of a formal statistical test justified by adequate theory ... " (Cohen, 1986, 112). Next, he bases his conclusion of stationarity after 1880 on estimated growth rates (log L) for the years from 1880 to 1960. For reasons not stated in his paper he ignored the years 1965, 1970 and 1975. I looked at these three years, and the result is that the downward trend in the growth rate becomes even clearer than one could infer already from Cohen's findings. In fact, the Swedish growth rate falls more or less steadily from 1815, with interruptions around 1880 (Cohen took that year as the start of the stationary period), and around 1940. It will be clear from this discussion that I underline Cohen's plea for formal inferential methods to test the assumptions of his method. Nevertheless, I assume that his method is inappropriate for industrialized countries, as many of these will have a demographic history similar to those of Sweden and the Netherlands.
Cohen's method is general enough to be applied not only to total population, but also to separate age-groups, provided that the stationarity assumptions hold. If the model is applied to a single age-group, recruitment of new cohorts takes place by means of "immigration".
The steadily declining growth rate in the Netherlands and Sweden is caused by falling birth rates. Hence the method probably cannot be used for the young age-groups. (In fact, because the fertility decline occurred between 1965 and 1975, for a population aged x, the stationarity assumption is unlikely in the period between 1965+x and 1975+x.) For other age groups the situation has to be tested. In section 8.3 I check the assumptions for age groups 0-4, 50-54 and 80-84. The young and the elderly displayed rather large errors in section 5.3, where I analysed the forecast uncertainty connected to the age structure of the population. The group 50-54 came out best in this investigation with errors less than 1 per cent.

8.3 | Age groups

The analysis will be restricted to the years from 1900 to 1989. Data for males and females in five-year age groups can be found in NCBS (1970) for the greater part of the period, and in the "Maandstatistiek van de Bevolking" for the recent decades. Growth rates were estimated according to expression (8.1), with jump-off year T equal to 1989, and base years 1900, 1901, 1902, ..., 1988.

As was expected, the pattern of the estimated growth rates for the age group 0-4 is that of a more or less regular decline from nearly 0.5 per cent per year around 1900 to -1.5 per cent in the mid-1970s, with interruptions around

1940 (due to a temporary increase in fertility which started in the mid-1930s) and around 1950 (due to the post-war babyboom). Next, the growth rate increases to nearly 2 per cent in 1988. The recent increase is caused by the echo of the post-war babyboom. As the long-term trend is definitely downwards, the stationarity assumption does not hold and point estimates and confidence intervals will not be calculated for this age-group.
Figure 8.2 shows estimated growth rates for males and females aged 50-54. A very clear decline in the growth rate is present in the period until 1960, whereas fluctuations in mortality and migration cause an unstable pattern in the years thereafter. First differences in W(t) suggest that a shift in the trend occurred around 1960 (figure 8.3). I assume that the series is stationary for the years 1960-1989, and that log(λ) and σ may be estimated from the data for these years. This leads to the estimates shown in table 8.2.

Figure 8.2. Estimated growth rates (log L) of population aged 50-54 in the Netherlands, jump-off year 1989, base years since 1900

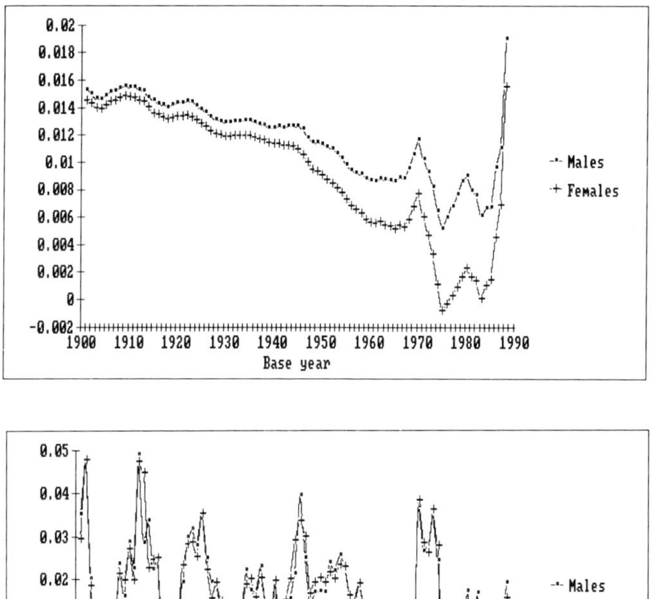

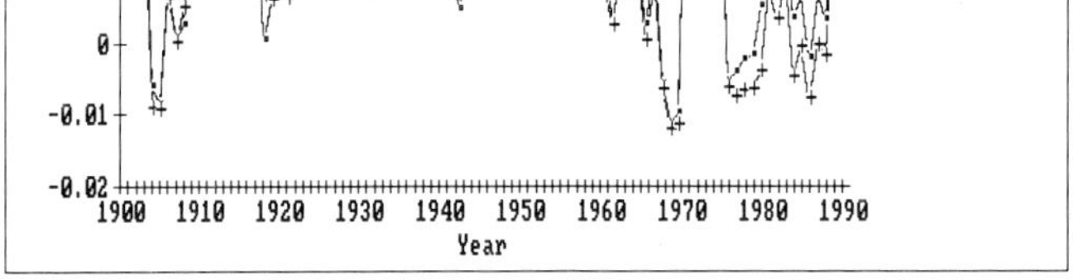

Figure 8.3. First differences in log of population aged 50-54, the Netherlands

Table 8.2. Estimates of growth rates (log(λ)) and of standard deviations of projection (σ) for the Netherlands population, computed according to (8.1) and (8.2), jump-off year 1989

	Age group			
	50-54[1]		80-84[2]	
	Males	Females	Males	Females
log(λ)	0.0087	0.0040	0.0208	0.0336
σ	0.0104	0.0166	0.0837	0.0972

Notes: 1) Data for the period 1960-1989.
2) Data for the period 1900-1989.

The age group 80-84 displays a slowly fluctuating growth rate since 1900. For males it moves between 1.5 and 2.5 per cent, for females mortality is lower and the rate lies between 2.5 and 4 per cent (figure 8.4). First differences in log of population exhibit irregularities due to mortality which seem to be less in recent decades than in the first half of this century (figure 8.5). Excess mortality for males in this age group started around 1960. Figures 8.4 and 8.5 suggest that log(λ) and σ may be estimated on the basis of the data for the whole period since 1900. Results are shown in table 8.2.

Table 8.3 presents projections according to expressions (8.3) and (8.4), with the estimates from table 8.2, and $z_{1/2\,\lambda}=1$. Hence, the intervals are 68 per cent confidence intervals. For the age group 80-84, the intervals are very wide, as could be expected: even with a probability as low as 68 per cent, the width is 60 (males) to 70 (females) per cent of the point estimate after 11 years. For durations beyond about 15 years the intervals become so wide that the projections are virtually no longer informative. It will not come as a surprise that with such large intervals, results for the age group 80-84 according to the medium variant of the 1989-forecast of the NCBS fall between the upper and the lower bounds given in table 8.3. More important, however, is that the high and the low variant of this forecast are much closer to each other than the upper and lower bounds of the confidence intervals: for 1995, for instance, the width is only 3 (females) and 4 (males) per cent of the value in the medium variant, and in 2000 it is only 5 (females) and 11 (males) per cent.

The results according to Cohen's method are much more precise for the age group 50-54. Even after 20 years the 68 per cent confidence interval is only 13 (males) and 20 (females) per cent of the point estimate. Yet the intervals are much wider than one would guess on the basis of the mean percentage

Figure 8.4. Estimated growth rates (log L) of population aged 80-84 in the Netherlands, jump-off year 1989, base years since 1900

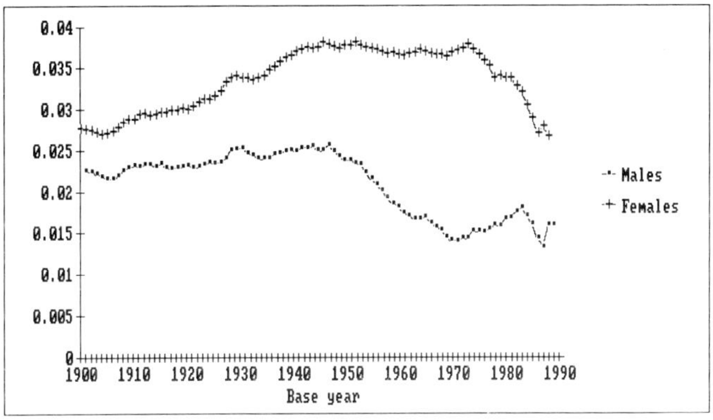

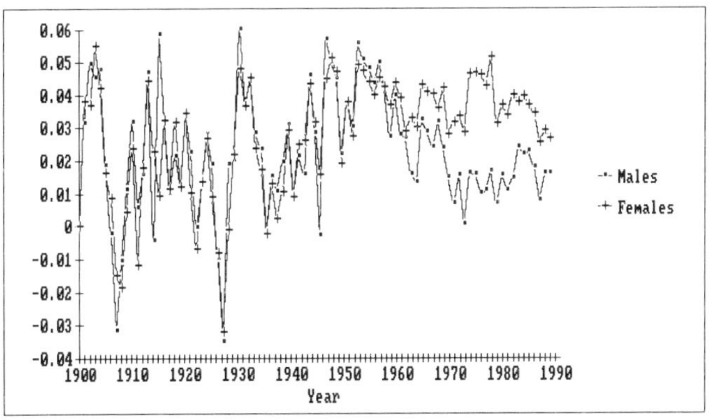

Figure 8.5. First differences in log of population aged 80-84, the Netherlands

errors and the root mean squared percentage errors which were calculated for historical forecasts - the latter measures do not exceed 1 per cent after 15 years for this age group, see section 5.3. Figure 8.2 offers an explanation for this discrepancy: the forecasts made between 1950 and 1970 could be based on rather stable growth rates for this age group. The projections in table 8.3 are an extrapolation of the relatively unstable trend in the period after 1960.
The ratio of the interval width of the age groups 80-84 to that of the age group 50-54 is between 5 and 8, depending on the length of the projection period (it also differs slightly by sex). This pattern agrees reasonably well with that for the ex-post errors in section 5.3.
The official forecast of 1989 shows a completely different pattern for the

Table 8.3. Projections for the Netherlands according to Cohen's method, age groups 50-54 and 80-84: point estimates and confidence intervals

	Age group 50-54 Males	Females	80-84 Males	Females
	in thousands			
1989	388	378	81	164
1990				
Point estimate	391	379	83	170
Upper bound	395	383	90	187
Lower bound	387	375	76	154
Interval width[1])	2	3	17	20
1995				
Point estimate	408	387	92	201
Upper bound	420	398	114	257
Lower bound	397	376	74	157
Interval width[1])	6	9	43	50
2000				
Point estimate	427	395	102	237
Upper bound	444	411	137	334
Lower bound	410	379	76	169
Interval width[1])	8	13	60	70
2005				
Point estimate	446	403	113	281
Upper bound	469	424	163	428
Lower bound	423	382	79	184
Interval width[1])	10	17	74	87
2010				
Point estimate	465	411	126	332
Upper bound	496	437	192	545
Lower bound	437	386	82	202
Interval width[1])	13	20	88	103

1) Difference between upper and lower bound, as a percentage of point estimate.

ages 50-54. The medium variant results in a strong rise until the year 2000 (581 thousand for males and 559 thousand for females), next a temporary decline (to 561 thousand for males and 546 thousand for females), and then an increase again (to 591 thousand and 580 thousand in 2020 for males and females). Moreover, the gap between the high variant and the low variant is much smaller: 5 thousand persons in the year 2020, and less in the earlier years.

Two reasons can be given for the divergence between Cohen's method and the cohort-component method. Firstly, one of the major features of the cohort-component method is that it can take due account of irregularities in the age structure. These are not included in the present application of Cohen's method. Secondly, Cohen's method rests on the assumption of stationarity in the growth rate, while the cohort-component method facilitates working with variable growth rates, for instance due to improved mortality conditions, or migration levels which deviate from those in the period prior to the jump-off year. So there is a certain trade-off between the two methods. With Cohen's method one obtains confidence intervals, and irregularities in the growth rate are reflected by wide intervals. With the cohort-component method, confidence intervals cannot be calculated, but one is able to compute point estimates (the medium variant) which may be more plausible than with Cohen's method. Irregularities in the growth rate may be decomposed into separate trends for fertility, mortality and migration, and these trends may be extrapolated.

9. CONCLUSIONS AND DISCUSSION

Suppes formulated a thesis about the incompleteness of scientific knowledge. " ... The naïve tendency is to reach at once for completeness and to be dismayed at the inability to achieve it. The proper attitude of empiricism, as I see it, is to expect incompleteness and to be surprised by its opposite ... Whether we are predicting the behavior of elementary particles, the weather, or international trade - indeed any phenomenon that has a reasonable degree of complexity - we can hope only to encompass a restricted part of the phenomena ... " (Suppes, 1984, 101, 115). This thesis applies particularly well to demographic phenomena such as childbearing, marriage, migration and the like. One has to accept the fact that our knowledge about such phenomena is very limited, and will remain limited in the future, too. Thus, predictions about the demographic behaviour of individuals are, and will be, uncertain. Errors are bound to arise in demographic forecasts. This study has looked into several aspects of uncertainty in the field of population forecasting at the national level: concepts, backgrounds, measurement, and empirics.

In this chapter I formulate and discuss the conclusions of this study, based on the results which were obtained in previous chapters. The main findings are summarized in section 9.1, in which I refer to the objectives and the research questions which were stated in chapter 1. A distinction is made between general findings, which immediately apply to national population forecasting in general (section 9.1.1), and empirical findings which were derived from the data that I collected for the Netherlands (sections 9.1.2 and 9.1.3). The latter empirical findings, which include the test results of the hypotheses, may only be generalized to other countries if the situation in these countries is thought to be comparable to that in the Netherlands. Section 9.2 contains some recommendations for the reduction of uncertainty in demographic forecasting at the national level, and suggestions how national statistical agencies may adequately deal with the uncertainty inherent to their forecasts. Finally, in section 9.3 I discuss some questions which are open to further research.

9.1 | Main findings

Five objectives of this study were formulated in chapter 1.
1 To describe the sources of the uncertainty which is connected with national demographic forecasts, and the reasons why it emerges.
2 To quantify that uncertainty by means of an evaluation of historical forecasts.
3 To explore possible regular patterns in ex-post observed forecast errors, and to attempt to link these errors to social, organizational and demographic background factors.
4 To give an estimate of the uncertainty to be expected in future population forecasts.
5 To give recommendations as to how the accuracy of future population forecasts may be enhanced.

In section 9.1.1 on general findings I discuss the first objective. The hypotheses which were formulated in chapter 3, and which were subsequently tested, relate mainly to the third objective. They will be investigated in section 9.1.2. The second and the fourth objective concern miscellaneous factors to be discussed in section 9.1.3, while the fifth objective regarding recommendations to statistical agencies will be taken up in section 9.2.

9.1.1. Generalities

* *Demographic forecasts are necessary*
Although this study has shown how uncertain demographic forecasts are, it is impossible to imagine a world without such forecasts. There is a scientific need for demographic forecasts, as there is a need for forecasts in any social science. Moreover, forecasting imposes a certain degree of rigour on thinking. And finally, the apparent need for demographic forecasts which exists for planners and policy makers implies that demographic forecasts will always be attempted by someone.

* *Three sources of uncertainty: randomness, ignorance, inherent unpredictability*
In chapter 1 I defined uncertainty as degree of forecast inaccuracy, and uncertainty was operationalized by forecast error. In other words, uncertainty was interpreted as a lack of predictability of the processes involved. I argued that there are three sources of uncertainty in national population forecasting: (i) randomness in population processes, (ii) ignorance of the processes involved, and (iii) inherent unpredictability of these processes. The first source arises from the fact that the point in time when individuals experience a certain event (e.g. childbearing, death, marriage, migration) is a random variable. The probability distribution (or rather, its parameters) can be inferred from observations on these events, and uncertainty (due to this first source) about the total number of events for a

group of (identical) individuals is large when the variance of the distribution is large, and/or when the number of individuals involved is small. Uncertainty due to randomness is the minimum amount of uncertainty one will encounter in a given situation; it is unavoidable. Uncertainty due to the second source, viz. ignorance of the processes involved, is, one hopes, only temporary. It arises because the probability distribution is unknown, and the answer to this second type of uncertainty is more research into human behaviour: one hopes to discover general behavioural laws which can be used for forecasting. Demographers who adhere to the second interpretation are much more optimistic than those who favour the third one: human behaviour is essentially unpredictable. Individuals have a variety of possible actions, which cannot be captured by one probability distribution (identical for all individuals concerned) for the time of the event. Thus population forecasts are inherently uncertain.

In practice one encounters all three types of uncertainty in a given situation. However, the record of population forecasting suggests that the third type of uncertainty (due to the inherent unpredictability of human behaviour) dominates the other two. Uncertainty due to randomness plays only a minor part in forecast errors. Attempts to discover more or less general behavioural theories, which can be applied fruitfully in population forecasting, have had only very limited success so far.

* *Demographic theories consist of conjectures, demographic laws express mere possibilities*

Most of the laws which make up demographic theories are in fact conjectures, that is statements of what *may* happen. They express relations between events or phenomena existentially - they are of the form 'if A, then sometimes B'. They are neither deterministic - 'if A, then always B' - nor probabilistic - 'if A, then most frequently B' -. The theories consisting of such conjectures are *local* and *partial*: they are not universally valid, rather they apply to particular cases only. Hence these theories have proved to be of only limited value to demographic forecasters. In practice, demographers resort to the subjective and naïve method of trend extrapolation. This explains why uncertainty is inextricably bound to population forecasting, and that demographic forecasts are inherently uncertain.

* *Errors in exogenous parameters are crucial*

In national forecasts of most industrialized countries, measurement errors, errors due to randomness of the processes involved, and errors because of inaccurate model specification are of only minor importance. The most common forecast errors arise because of a misspecification of the values of exogenous parameters, that is the total fertility rate (or completed cohort fertility), the life expectancy at various ages, and the numbers of immigrants and emigrants.

* *The cohort-component method is the standard for national forecasts - crucial stages*
  Although other approaches exist, e.g. economic-demographic models, time-series models, microsimulation models and judgemental approaches, the cohort-component method is the standard method which is used by statistical agencies for demographic forecasting at the national level. This method is entirely deterministic, and the model involved is of the accounting type - no behavioural relationships are involved.
  Considered from the accuracy point of view, there are three crucial phases in the process of constructing a population forecast. First, subjective decisions have to be made at the time when the forecaster is confronted with the choice whether or not to include other demographic characteristics than age and sex in the forecast, such as region of residence, marital status, nationality etc. Second, subjectivity is abundant when it comes to the interpretation of historical trends, and a judgement on the relevance of non-demographic variables for the explanation of these trends. And finally, parameter extrapolation is the most subjective of all forecasting tasks. It adds to the subjectivity of the previous tasks the fact that the forecaster must now decide whether relationships assessed for the past will also hold in the future. Quantitative models that explain demographic behaviour cannot be used here, for reasons just discussed.

* *Instruments to express uncertainty in traditional forecasting: variants, target years, presentation*
  In spite of the deterministic character of the traditional cohort-component approach for demographic forecasting, a number of instruments exist to express forecast uncertainty.
  Variants are an effective way of expressing the uncertainty which is introduced when qualitative assumptions regarding the most likely future paths for fertility, mortality and migration have to be quantified into values of such indicators as the total fertility rate, the life expectancy and the number of migrants. For a particular indicator in a certain year, the distance between its values in the high variant and the low variant is indicative of the degree of uncertainty the forecaster wishes to attach to that indicator. Thus variants are to be interpreted as uncertainty variants. Some demographers give a different interpretation to the notion of forecast variants, viz. that of alternative futures. However, this interpretation is inconsistent with the accepted definition of a forecast: if variants are interpreted as alternative futures the question remains which variant sketches the most likely future.
  As we look further into the future, uncertainty grows, and the range between high and low variant should grow accordingly. Moreover, there are large differences among components in range between variants.
  A *target year* (or target cohort) marks, for a specific demographic indicator, the approximate end of the period in the future for which the forecaster is able to select, from a large range of possible future trends, a plausible

or likely path. The distance between the jump-off year of the forecast and the target year expresses the degree of uncertainty for the component concerned. Because the forecaster is unable to indicate, for a particular component, the plausibility of its trajectory beyond the target year, a forecast gradually turns into a mere projection as one considers periods around and beyond that target year.

A proper *presentation* is of great importance for an effective communication of forecast uncertainty to the user. Forecast variants should be given equal weight in the presentation, for instance, by reserving, whenever possible, one dimension in any table of forecast results for the variants actually used. Placing forecast results in a historical context can also give the user an appropriate impression of the uncertainty involved, because the greater the uncertainty, the more often and the more drastically the forecast has to be adjusted. Each forecast report ought to include a separate section in which an account is given of the accuracy of a series of previous forecasts (not just the most recent).

9.1.2. Tests of hypotheses

In chapter 3, sixteen hypotheses about accuracy and errors in national demographic forecasts were formulated. As I explained in chapter 7, not all these hypotheses could be tested with the data that I have for the Netherlands. For several reasons which are set out in section 7.1.2, hypotheses numbered as 5, 6, 8, 11, 13, 14 and 15 (see section 3.4 for these numbers, as well as the numbers of other hypotheses used in this chapter) had to be disregarded. They will be considered in section 9.3. The findings regarding the remaining nine hypotheses are discussed below.

1 Irregularities increase forecast errors

The slope (upward or downward) of observed fertility trends (hypothesis 1.a) increases errors in birth rates, see estimation results for models 7.1-7.4. The coefficient of SY varies between 0.2 and 0.3, which means that a rise or fall in the crude birth rate by 1 per thousand over a period of five years causes an error in the birth rate in the middle of that period of 0.2-0.3 per thousand.

Hypothesis 1.a was not confirmed for mortality or international migration.

Fluctuations in observed data (hypothesis 1.b) have an ambiguous effect on forecast errors. Large fluctuations increase forecast errors for birth rates in the short run, as well as errors in mortality forecasts for females. However, this hypothesis could not be confirmed for male mortality, or for long-term fertility errors.

Current levels of key demographic indicators dictate to a large extent their extrapolated values (hypothesis 1.c, cf. also the discussion on "assumption drag" in section 9.1.3). For *fertility* this is illustrated very clearly in figure 9.1, which gives observed average numbers of children (age-specific and marital duration-specific) by calendar years, and target fertility values used

in forecasts with corresponding jump-off years. It is surprising to see how close the target values are for the forecasts of 1970, 1972 and 1975 (indicated with a + sign) to the observed level of fertility in the jump-off year (see dotted line). Similarly, the * signs, indicating target values in forecasts made between 1980 and 1986 correspond very closely to observed TFR's (solid line). In fact, target values are so close to observed values in the jump-off year that it looks as if no explicit extrapolations were carried out on the basis of historical trends, but that, instead, for each forecast the most recently observed values were kept constant.
Strong evidence which supports hypothesis 1.c for the case of *mortality* is to be found in figures 5.6 and 5.7, see also tables 7.1.a and 7.1.b.

Figure 9.1 Average number of children, the Netherlands, observed and target values

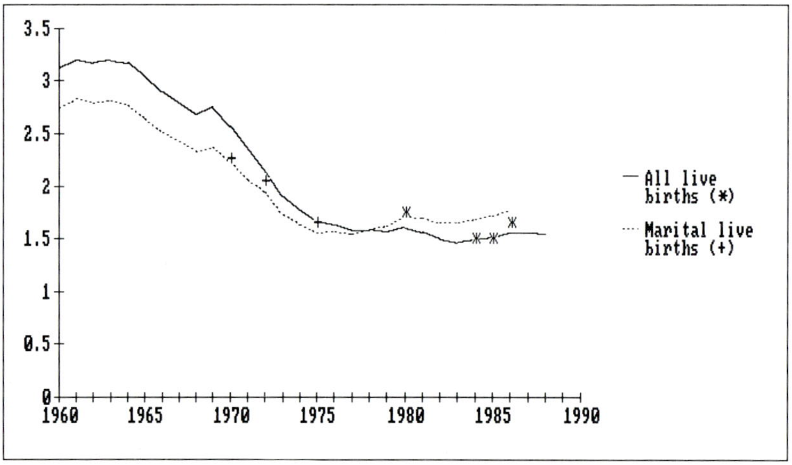

Note: For forecasts with two variants, the average of the two target values was taken. For forecasts with three variants, the target value of the medium variant was used.

2 Ex-post observed population trends affect forecast accuracy

Because forecast accuracy has been operationalized here using the concept of forecast error, this statement may seem a tautology. The important question is, of course, to what extent accuracy is affected by observed developments, and to what extent by other factors. In chapter 5 (section 5.4.1) the errors in marital fertility were decomposed into a fertility component and a nuptiality component. This analysis resulted in strong evidence in favour of the hypothesis that there is a very strong impact of observed trends on accuracy: during the years with a sharp decline in fertility (1965-1975), this component explained almost entirely the errors in marital fertility; when, after the mid-1970s, marriage was delayed, or even substituted for unmarried cohabitation, marriage rates declined and errors in marital fertility were largely affected by nuptiality.

3. A forecast's jump-off year is strongly correlated with forecast errors

At several levels of analysis, strong support was found for this hypothesis. Errors were investigated in a number of demographic variables: total population growth (section 5.2), absolute numbers of births and deaths, crude birth rates and crude death rates, summary indicators such as average number of marital live births, and life expectancy at birth (section 5.4). Corrections were made for possible confounding effects due to length of forecast period (LOFP) and calendar year (chapter 6), and also for the irregularities in observed trends and the autoregressive character of errors in birth rates and life expectancies (chapter 7). Even for the most detailed analyses, the jump-off years 1965, 1967 and 1970 displayed a considerable impact on fertility errors, see the estimates in expression (7.1'). For mortality of both males and females, the jump-off years 1950, 1951, 1970, 1972 and 1975 correlated strongly with errors in life expectancy, see expressions (7.4) and (7.7).

4 The assumption that major changes in forecast methodology in the Netherlands have resulted in more accurate forecasts is not supported by the data

This statement is formulated more cautiously than the others. The reason is that the variables which were used to operationalize forecast effect, that is the average error which is unique for a certain forecast, include more than methodology alone. For instance, a (possible effect due to a) change in the organizational setting in which the forecast is produced is captured by these variables as well. Nevertheless, the major difference between the 1967-forecast and that of 1970 was the new methodology to analyse and extrapolate fertility: the NCBS changed from an age-specific period approach to a marriage duration-specific longitudinal approach. If one corrects for possible distortions due to LOFP-effects or calendar year effects, the average fertility error in the 1970-forecast is not significantly smaller than that in the 1970-forecast, see tables 6.3 and 6.4. Similarly, despite corrections for irregularities in observed trends and for the autoregressive character of fertility errors, expression (7.1') indicates no improvement between 1967 and 1970.
A second change in forecast approach took place around 1975, when more attention was given to the sociodemographic background of key parameters, and when organizational changes were carried through. The separation model of chapter 6 provides estimates which indicate a slight improvement in the accuracy of fertility and mortality forecasts around 1975, but for fertility about one-fifth of this effect disappears after one corrects for the fact

that births followed a relatively smooth trend after 1975. For mortality (for which irregularities have no impact on forecast errors, see the discussion of hypothesis no. 1) the change in forecast approach was followed a few years later by increased accuracy: the estimates of average accuracy of the 1980-forecast for life expectancy are lower than those of the 1975-forecast, and substantially below those of the forecast of 1972, see the coefficients of D72, D75 and D80 in expressions 7.4 and 7.7.

7 Too short base periods lead to large errors in mortality forecasts

The number of years contained in the analysis of observed trends before the extrapolations are made affects the accuracy of mortality forecasts. Forecasts for which the base period was only five years had errors in the life expectancy which were 0.6 to 1.2 years higher than those with a base period of ten years, see expressions (7.6), (7.9) and (7.10). It should be realized, however, that errors arising because of too short base periods are much smaller than those caused by other factors, such as the length of the forecast period.
An important question is, of course, what the optimum length of the base period is in a given situation. To answer that question would require a much more detailed analysis than could be given here. The number of years which were used to analyse observed mortality varies from three (the 1950-forecast) to 14 (the forecasts of 1975 and 1980), with an average of 8.4 years. It would seem good practice to use a period of 8 to 14 years.

9 Forecast results become more uncertain the further we look into the future; uncertainty increases approximately according to a linear or a square-root function of the length of the forecast period

The descriptive analysis in chapter 5 revealed that fertility errors rise more or less linearly as the length of the forecast period (LOFP) increases - the slope is approximately 7-8 percentages points per five years for the mean percentage error, and 10-12 points for the root mean squared percentage error (see section 5.5.2). Confounding effects of calendar year and forecast jump-off year may have blurred the picture. The more detailed separation analysis of errors in numbers of births in chapter 6 resulted in duration patterns for which a linear form performs as well as a square-root form. An exponential form fits the data badly.
For mortality, the increase is slower than that displayed by a straight line. Errors level off after a duration of approximately 15 years (see section 5.5.2).
Signed errors in crude birth rates and in crude death rates do not exhibit a regular duration pattern. The reason was given in sections 6.3.1 and 6.3.2: in the past 40 years, population growth was stronger than the growth of errors in numbers of births and deaths. In other words, duration patterns in errors of births and deaths forecasts (absolute numbers) are caused by rela-

tively strong population growth. In times when growth is absent, no consistent duration patterns are found in these errors. Indirectly these findings also lend some support to hypothesis no. 14: population growth tends to increase forecast errors. According to the accepted hypothesis no. 9, the mechanism runs through forecast duration.
Contrary to the finding for signed errors, duration patterns *are* consistent for *unsigned* errors: absolute errors in birth rates and in life expectancies grow according to a square-root function of duration, with coefficients of about 1.4-1.7 for fertility, and 0.4-0.5 for mortality. Thus for mortality we can look further into the future than for fertility. After 15 years, errors in mortality forecasts are smaller than those which arise in fertility forecasts after only five years.

10 The best predictor of the forecast error in a certain year is the error for that particular forecast one year previously

At least 92 per cent of the variation in the long-term error for birth rates and life expectancies in a particular year was "explained" by the corresponding error in the preceding year, and 83 per cent of the short-term error. Thus the direction of the initial take-off for the forecast and real trend is of decisive nature for the errors. The error observed in the first year after the forecast was produced is a reliable indicator for the error pattern in the following years.

12 Errors are highest when behaviourally determined demographic variables are involved

Strong support for this hypothesis was found by comparing errors in fertility to those in mortality. In general, mortality errors are only 20 to 45 per cent of those in fertility. In many cases the ratio is 1:3. Evidence can be found in tables 5.8 and 5.11 by comparing "Total" columns for different periods; in tables 6.3 and 6.5 by looking at estimated coefficients of k and $\sqrt{k}$; and in expressions (7.3), (7.6), (7.9), (7.10), (7.14), (7.18), (7.19), and (7.20) - see the estimates for the SQRLP-coefficients.

16 There is no evidence for a continuous increase in forecast accuracy due to autonomous technological change

One would expect that such factors as the steadily increasing body of scientific knowledge and of available data, continuous improvements in computer hardware and software, intensified international cooperation, etc. improve forecast accuracy. No evidence was found for this hypothesis. Fertility forecasts show a more or less steady improvement after 1965 (see, for instance, tables 6.3 and 6.4), but this effect disappears if overestimations count as much as underestimations, and if, moreover, one controls for irregularities in ex-post observed trends and the autoregressive character of the error itself

(see the estimates in expression (7.1')). In that case the accuracy of the forecasts of 1965, 1967 and 1970 is far below average, while the remaining forecasts just show small errors (I discuss this issue further under the item "Assumption drag", see section 9.1.3).
Mortality errors were high in the forecasts of 1950, 1951, 1970, 1972 and 1975, and much smaller in the forecasts of 1956, 1965, 1967 and 1980, see table 6.5 and estimates in expressions (7.4) and (7.7).
Thus instead of a steady improvement in forecast accuracy, one sees rather strong fluctuations in quality.

9.1.3. Miscellaneous empirical findings

* *Evaluation of historical forecasts*

In chapter 5 I carried out a descriptive analysis of errors in national demographic forecasts, produced by the Netherlands Central Bureau of Statistics (NCBS) since 1950. The investigations in that chapter were motivated by the following research questions. How large are the errors in the results (output) and the assumptions (input) of national population forecasts? What is the accuracy of stock variables in a forecast, such as the total population size or the population broken down by age and sex? And what is the accuracy of flow variables: numbers of live births, deaths and migrants? How does the accuracy of these forecast results compare to that of indicators which are used to set up the assumptions: total fertility rate, life expectancy etc.?
The findings of that chapter can be summarized as follows.
The errors in age structures exhibit a very regular pattern, which is characteristic for a country with recent strong drops in fertility, and a steadily improving mortality. The pattern is that of large errors at young and high ages, and only minor errors (that is, when the LOFP is not more than 15 years) for age-groups between 20 and 75 years.
Concentrating on fertility, an analysis of errors in annual numbers of births, in crude birth rates, and mean numbers of children reveals that the error patterns for these three indicators are almost identical for the short term and the medium term (LOFP up to 15 years). On the long term, the three patterns display very irregular trends.
For mortality, errors in life expectancies are much smaller and much more regular than those in annual numbers of deaths or crude death rates. The error levels in numbers or rates are two to three times higher than those in life expectancies. Thus a large part of the error in mortality results is due to errors in population size and age structure.
International migration was only introduced in Dutch national forecasts in 1975, and the analyses of error patterns for this component could only be very rudimentary. Yet it was found that the formal introduction since 1975 of external migration as a component of change in Dutch forecasts generally improved their accuracy, although an assumption of zero net-immigration implicitly contained in the forecasts produced in the 1950s

did not yield less accurate results. The relatively stable trends in actual emigration resulted in more accurate forecasts for that component than those for immigration, for which the low degree of accuracy in the Netherlands is primarily due to irregular inflows of migrants from non-European countries.
Mortality errors are much lower than those for fertility, and the gap widens as the LOFP increases. For instance, after five years, mortality errors are 40 to 44 per cent of those in fertility. After ten years, the ratio has diminished to 33-37 per cent, and after fifteen years it is only 21-29 per cent.

* *Confidence intervals*
In chapter 8, attempts were carried out to construct confidence intervals around forecast results. Considerable work has still to be done here. For total population size, the method of Stoto works well. For the age structure, the method suggested by Cohen is inappropriate for many age groups: its condition of (stochastic) stationarity in observed matrices of transition probabilities in a certain period prior to the jump-off year will be violated in many industrialized countries which experienced a fertility decline in the 1960s. For age groups for which stationarity is not an unreasonable assumption (50-54 and 80-84 in the present application, and probably also for many of the intermediate age groups; forecast period up to 20 years), confidence intervals are wider than one would expect on the basis of historical forecasts. The relative age-specific pattern (wide intervals for high age groups, narrow ones for age groups around 50) is the same in Cohen's method as with ex-post errors.
Other methods than those of Stoto and Cohen have to be tested in practical situations, particularly for the purpose of constructing confidence intervals around age structures. The approach of Williams and Goodman (1971) is an empirical method, which can be applied when a large number of forecasts are available. Alho and Spencer (1985) specify a model in which the vital rates evolve through time according to low-degree polynomials. Their method rests on a number of assumptions (e.g. about the autocorrelations and cross-correlations of age-specific vital processes), some of which are difficult to test. A detailed study could probably assess the merits of this method.

* *Assumption drag*
The discussion of hypothesis 1.c in this chapter illustrated very clearly how severely forecasters are affected by an "assumption drag": they continue to use (most often unintentionally) assumptions which already have been refuted by the data for some time. The reason why forecasters lag behind actual trends is that population forecasting is based *not* on valid models for the demographic behaviour of the population, but on the extrapolation of regularities observed in actual trends, see the discussions in chapter 3. Consequently, the assumption drag arises because temporary

fluctuations are frequently difficult to distinguish from the start of a trend shift. To separate these two, data is often required over quite a long period. Moreover, it takes some time to prepare a demographic forecast, and thus initially valid assumptions may be outdated by the time the forecast is published (Ascher, 1978, 138).

One of the consequences of this assumption drag is an overreaction of the forecaster when it is noted that earlier forecasts lagged behind actual developments. Examples for fertility can be found in the results of models 3 and 4 of table 6.4, and in expression (7.1). Firstly, forecast effects follow approximately an alternating pattern around a general trend, see, for instance, the forecast effects for the forecasts of 1965-1980 in table 6.4, and the coefficient estimates for the dummy variables of those forecasts in expression (7.1). Secondly, the large errors in the fertility forecasts of 1965-1972 induced a margin between the high variant and the low variant of the 1975-forecast which was about 25 per cent wider than the average margin. The margin between the high and the low variant of the 1980-forecast was rather narrow, thus reflecting the optimism among the NCBS-forecasters about the predictability of births after a period of stable fertility trends since 1975. For mortality I discussed reaction patterns in subsequent forecasts in connection with table 5.12 and figure 5.7: the mortality assumptions of the 1970-forecast can now be considered a somewhat too strong reaction to the underestimations of male and female life expectancy in the 1965-forecast.

In section 9.2 I propose two solutions for the assumption drag phenomenon: an interdisciplinary team of forecasters, and frequent forecast production.

* *A qualifier*

Cross-disciplinary comparisons of the accuracy of forecasts have shown that *within the social sciences*, demographic forecasts do not perform badly (Ascher, 1978; Becker and Dewulf, 1990). Forecasters in other areas, such as the educational system (pupils, students, staff), socio-cultural issues (e.g. female emancipation), the labour market and election polls are in a relatively unfavourable situation. They have less data, less insight in behavioural processes, and less refined forecast instruments than demographers. In these respects, demographic forecasters are close to their colleagues in such fields as economy, transportation, meteorology and energy. A typical distinction between forecasts from various disciplines is, of course, the speed with which underlying processes develop: this ranges from a period of a few days in weather forecasting to a century or so in predictions about civilizations.

9.2 | Recommendations for current population forecasting

Based on the results of this study, in combination with findings reported in the literature, I now formulate a number of recommendations for an adequate handling of uncertainty and a reduction of errors demographic forecasts.

* *A demographic forecast should be prepared in an interdisciplinary team*
Because national population forecasts are produced with a naïve and subjective method, the potential dangers of an assumption drag can be avoided by working with a team in which, apart from demography, disciplines such as sociology, psychology, history, economics and statistics are represented. In the assumption-making process, each discipline should have a balanced input. Thus, because objectivity is unattainable, intersubjectivity should be strived for. Final decisions regarding the forecast assumptions should not be made on the basis of hierarchical considerations. Delphi-like decision-making processes are worth considering.

* *Forecasts should be produced frequently*
The assumption drag leads to overreactions on earlier major misses. Annual or bi-annual forecasts reduce the risk of such overreactions. Moreover, frequent forecasts make the user aware of the uncertainty involved in demographic processes. Population variables which often have to be updated are, by implication, uncertain. Unpleasant surprises for the user can be avoided, and the user is forced to take uncertainty into account.

* *Uncertainty should be reflected in the presentation of forecast results*
In section 2.3.3 I reviewed a number of possibilities the forecaster has for communicating uncertainty to the user. Variants should be given equal weight in the presentation: they should be genuine "uncertainty variants". Alternative futures may be relevant, but it should be stated very clearly which of the alternatives is (are) considered realistic. Due attention should be given in the forecast report to the ex-post accuracy of previous forecasts. An indication should be given of the consequences of a situation in which historical errors would also apply to the present forecast. Comparisons should be presented of key forecast results (e.g. the population structure in the year 2020) in a series of forecasts produced in the previous few decades.

* *Variants, target years and target values*
Forecasters should carefully consider, for each component, whether or not to include variant assumptions. Aspects which are important here were listed in section 2.3.1: the notion of uncertainty variants, the significance of each individual component for the forecast results, and irregularities in the past course of the components. Differences in the degree of uncertainty between components may be reflected in two ways: (i) the gap between corresponding variables in the high and the low variant (which may be set on the basis of ex-post observed errors), and (ii) the length of the period between the jump-off year and the target year, that is, the last year of the

(explicit) extrapolation period. Calculations for years beyond the target year(s) based on constant levels of input parameters are mere projections, not forecasts.

* *The separation model in chapter 6 offers a tool for setting priorities in research in the field of demographic forecasting*
If forecast errors turn out to be influenced mainly by real trends in demographic indicators that are observed after the forecast was produced, and much less so by a choice for more advanced forecast methodologies, then monitoring the actual trends should have a much higher priority for forecasters than the development of more sophisticated models and methods. The separation model in chapter 6 identifies period effects and forecast effects in errors obtained from a series of historical forecasts. The period effects reflect social and demographic conditions observed during a certain period in time after the forecast was produced, and the consequences these conditions may have had on forecast quality. The forecast effect is seen as an indicator which expresses, in one number, the average error connected to that forecast, irrespective of period in which the error is observed. For the Netherlands, period effects and forecast effects in fertility and mortality are approximately of the same order of magnitude, see section 6.3.4. Thus, monitoring (ex-post) observed trends for these two phenomena should be given approximately the same priority as efforts to improve the conditions under which the forecast is produced, and its methodology. For international migration, period effects are somewhat larger than forecast effects, see table 6.6. Thus, a considerable part of migration errors cannot be avoided.

9.3. | Open research questions

A variety of issues are worth further consideration.
First of all, six of the sixteen hypotheses formulated in chapter 3 were not tested, for reasons set out in chapter 7. The issue whether experts have a positive impact on forecast accuracy, relative to mechanical approaches (hypothesis no. 5) could be investigated if one had data for a country which shifted from an approach in which all relevant input parameters were kept constant, to an approach with explicit extrapolations of (at least key) input parameters on the basis of judgemental considerations. The possible effects of a change in bureaucratic decision-making (hypothesis no. 6) and of the type of data used in the extrapolation (hypothesis no. 8) also require better data than in the present study. Hypothesis no. 11 states that a combination of forecasts results in smaller errors than each of the constituent forecasts. It would be very difficult to test this hypothesis if a cohort-component methodology were to be combined with a forecast produced with an entirely different approach. The reason is that the cohort-component approach is so firm a standard, that alternative forecasts will be constructed perhaps occa-

sionally, but probably not on a regular basis. However, if one considers this hypothesis from the input parameter perspective, a combination of different approaches automatically results when working with an interdisciplinary team, see the discussions above: no single discipline has had better insights in future demographic patterns than other disciplines, and thus a combination of disciplines is assumed to be advantageous. A case study could provide the data to test the hypothesis in this form. The hypothesis that population size correlates negatively with forecast errors (no. 13), although not tested in the present study, has been verified on a number of occasions, see the references listed in section 3.4. In hypothesis no. 14 it is assumed that population growth increases forecast errors. Indirect arguments which support this hypothesis are to be found in section 9.1.2 in the discussion concerning hypothesis no. 9. Finally, hypothesis no. 15 states that the frequency of measurement in the base period, and the length of the unit forecast interval, may be related to forecast accuracy. To test this assumption would require data from a country in which forecasters work with an interval of five years, say. The accuracy of these forecasts should be compared to those of a country in which forecasts are made with the more usual interval of one year. For both countries, a single forecast would not suffice. Inter-country differences in actual demographic developments should be corrected for, and this is possible when a separation model similar to that of chapter 6 is applied to errors in a series of forecasts of both countries.

Secondly, considerable work remains to be done before appropriate confidence intervals around forecast results can be constructed. This issue was discussed in section 9.1.3.

Thirdly, social psychology could reveal important insights in the assumption-making process of population forecasters. This process was typified earlier as naïve and subjective. It would be of great relevance to understand on what basis an individual forecaster selects certain trajectories for key demographic indicators as more likely than others. Is it possible to make the decision rules explicit? To what extent does the individual forecaster base his decisions on demographic and non-demographic information? How important are the opinions of other relevant individuals? What is the impact of the forecaster's own demographic situation (age, health, household type, family size) and history? How does he react to major errors in earlier forecasts in which he was involved - and in earlier forecasts in which he was *not* involved? I assume that a proper understanding of the mechanisms (if any) of human thinking relevant to population forecasting is as important for forecast accuracy as the factors which were investigated in the present study.

A final issue to be addressed is the following. In section 1.3.2 I referred to Suppes (1984, 31) who stated that the real test of determinism is predictability. A recent new field of science, which is known under such diverse names as "chaos mathematics", or "non-linear deterministic systems", challenges this opinion, and it may have great potential for the study of the predictability of population processes. The idea, which stems from weather forecasting, is that reality is inherently non-linear. Some non-linear systems, such as those with underlying differential equations of the logistic type, behave erratically in certain critical areas of their parameter space. Such systems may display stable equilibrium behaviour, but once their parameters have surpassed so-called bifurcation points, their behaviour becomes essentially unpredictable. It is not a possible random component in the model that causes unpredictability, but the non-linear nature of these deterministic systems. A weaker form is that of sustained cyclical behaviour, or "limit cycles" (see, for instance, Frauenthal and Swick, 1983; Feichtinger and Sorger, 1990; Wachter and Lee, 1989).

The field of non-linear deterministic systems is relatively new in demography. Land (1986, 897) and Land and Schneider (1987, 17) discuss some of the aspects involved. Bonneuil (1989) gives a first application to the rapid fertility decline in nine European countries in the 1960s. A major breakthrough in the study of uncertainty in population could be expected if one could demonstrate that the population system is non-linear, and that trend shifts in key demographic indicators in the past are the result of small changes in one or more of the parameters of that system.

REFERENCES

Ahlburg, D.A. (1982) How accurate are the U.S. Bureau of the Census projections of total live births? *Journal of Forecasting* 1, 365-374.

Aldridge, J.H. and F.D. Nelson (1987) *Linear probability, logit, and probit models*, Sage University Paper Series on Quantitative Applications in the Social Sciences no. 07-045, Beverly Hills and London, Sage Publications (fourth printing).

Alho, J.M. (1984) Probabilistic forecasts - the case of population projections, *Scandinavian Housing and Planning Research* 1, 99-105.

Alho, J. (1985) Statistical problems in demographic forecasting, paper PAA Annual Meeting Boston, March 1985.

Alho, J.M. (1988) Stochastic methods in population forecasting, unpublished paper Institute for Environmental Studies and Department of Statistics, University of Illinois at Urbana-Champaign.

Alho, J.M. and B.D. Spencer (1985) Uncertain population forecasting, *Journal of the American Statistical Association* 80(390), 306-314.

Alho, J.M. and B.D. Spencer (1988) Stochastic population forecasting, paper 8th International Symposium on Forecasting, Amsterdam, June 12-15, 1988.

Amemiya, T. (1985) *Advanced Econometrics*, Oxford, Basil Blackwell.

Armstrong, J.S. (1985) *Long range forecasting: From crystall ball to computer*, J. Wiley, New York (2nd edition).

Arthur, W.B. (1984) The analysis of linkages in demographic theory, *Demography* 21(1), 109-128.

Ascher, W. (1978) *Forecasting: an appraisal for policy-makers and planners*, The Johns Hopkins University Press, Baltimore and London.

Ascher, W. (1988) Beyond accuracy: progress and appraisal in long-range political economic forecasting, paper 8th International Symposium on Forecasting, Amsterdam, June 12-15, 1988.

Bandel, J. (1990) Befolkningsprognoser - hur säkra är de? Utredningsrapport nr. 1990:1, Stockholm Office of Research and Statistics.

Bandel, J. and S. Schéele (1990) Population forecasts - how (un)certain are they? Paper 17th International Conference on Regional and Urban Statistics, Duisburg, June 1990.

Bartlema J. (1987) Developments in kinship support networks for the aged in the Netherlands: A social-demographic analysis, Catholic University of Brabant, Tilburg, the Netherlands (Sociale Zekerheidswetenschap Rapporten no. 3).

Baumgartner, Th. (1987) Population forecasting, policy-making and socio-political context, Paper European Population Conference Jyväskylä, Finland, June 11-16.

Baumgartner, Th. and A. Midttun (1987) *The politics of energy forecasting: A comparative study of energy forecasting in Western Europe and North America*, Clarendon Press, Oxford.

Baxter, R. and I. Williams (1978) Population forecasting and uncertainty at the national and local scale, *Progress in Planning* 9, 1-72.

Baydar, N. (1983) Analysis of the temporal stability of migration patterns in the context of multiregional forecasting, NIDI-Working Paper no. 38, Netherlands Interuniversity Demographic Institute, Voorburg, the Netherlands.

Becker, H.A. and G. Dewulf (1990) Ontwikkeling en kenmerken van toekomstonderzoek en zijn evaluatie, in: H.A. Becker and G. Dewulf (eds.) *Terugkijken op toekomstonderzoek*, ISOR/ University of Utrecht, 1-37.

Beets, G.C.N. and A. de Graaf (1980) Onderzoek Gezinsvorming september 1977, *Maandstatistiek van Bevolking en Volksgezondheid* 23(12), 41-60.

Beets, G.C.N. and H.G.J.M. Cruijsen (1985) Bevolkingsprognose 1984: hypothesen betreffende de nuptialiteit, *Maandstatistiek van de Bevolking* 33(10), 50-68.

Beets, G.C.N. and C.J.M. Prins (1985) Bevolkingsprognose 1985: hypothesen betreffende de sterfte, *Maandstatistiek van de Bevolking* 33(7), 48-60.

Beex, B. (1980) Niet-eerste huwelijkssluiting in Nederland: Verleden, heden en toekomst, NIDI-Intern Rapport no. 21, Netherlands Interuniversity Demographic Institute, Voorburg, the Netherlands.

Berk, R.A. and T.F. Cooley (1987) Errors in forecasting social phenomena, *Climatic Change* 11, 247-265.

Birg, H. and R. Brock (1986) Zur stochastischen Variabilität von Bevölkerungsprognosen für Regionen bzw. für Bevölkerungsgruppen - Vergleich alternativer Simulationsmodelle an Fallbeispiel der Stadt Werne. Universität Bielefeld, Institut für Bevölkerungsforschung und Sozialpolitik (IBS Materialien nr. 22).

Bishop, Y.M.M., S.E. Fienberg and P.W. Holland (1975) *Discrete Multivariate Analysis: Theory and Practice*, The MIT Press, Cambridge (Mass.).

Bonneuil, N. (1989) Conjoncture et structure dans le comportement de fécondité, *Population* 44(1), 135-157.

Boudon, R. (1986) *Theories of social change: A critical appraisal*, Polity Press, Cambridge.

Brass, W. (1974) Perspectives in population prediction - illustrated by the Statistics of England and Wales, *Journal of the Royal Statistical Society* A 137(4), 532-583.

Brass, W. (1985) The potential usefulness of demography, *International Population Conference Florence* 1985, 5-12 June, Vol. 1 (IUSSP, Liège), 33-46.

Brass, W. (1986) Uses of population projections, paper presented at the Conference on Population Research in Britain, University of East Anglia, 10-12 September 1986.

Bretz, M. (1986) Bevölkerungsvorausberechnungen: Statistische Grundlagen und Probleme, *Wirtschaft und Statistik* 4, 233-260.

Brunborg, H. (1984) Hvor sikre er befolkningsprognosene? Noen prinsipielle betraktninger om usikkerhet i befolkningsprognoser, Tekniske Rapporter no. 37, Nordisk Sekretariat, Copenhagen.

Bunn, D.W. (1988) Combining forecasts, *European Journal of Operational Research* 33(3), 223-229.

Capon, N. and J.M. Hulbert (1985) The integration of forecasting and strategic planning, *International Journal of Forecasting* 1, 123-133.

Carbone, R., A. Andersen, Y. Corriveau and P.P. Corson (1983) Comparing for different time series methods the value of technical expertise, individualized analysis, and judgemental adjustment, *Management Science* 29(5), 559-566.

Carter, L.R. and R.D. Lee (1986) Joint forecasts of U.S. marital fertility, nuptiality, births and marriages using time series models, *Journal of the American Statistical Association* 81, 902-911.

Churchhouse, R. (1985) Uncertainty in mathematics and computing, in: S. Nash (ed.) *Science and Uncertainty*, Proceedings of a conference held under the auspices of IBM United Kingdom Ltd., London, March 1984, Science Reviews Ltd., 87-91.

Clogg, C.C. (1982) Cohort analysis of recent trends in labour force participation, *Demography* 19, 459-480.

Cohen, J.E. (1985) Mathematical demography: Recent developments in population projections, paper presented at the IUSSP International Population Conference, Florence, June 1985.

Cohen, J.E. (1986) Population forecasts and confidence intervals for Sweden: a comparison of model-based and empirical approaches, *Demography* 23(1), 105-126.

Condon, K. and T. Pullum (1988) Underestimation in the United States Census 1880-1900: An application of Age-Period-Cohort Analysis, Paper PAA Annual meeting New Orleans, LA, 21-23 April, 1988.

Cruijsen, H.G.J.M. (1985) Prognose van de Bevolking van Nederland na 1984, *Maandstatistiek van de Bevolking* 33(4), 30-43.

Cruijsen, H.G.J.M. and W.D. van Hoorn (1983a) De waargenomen en geprognosticeerde bevolkingsontwikkeling in Nederland 1980-1982, *Maandstatistiek van de Bevolking* 31(8), 16-23.

Cruijsen, H.G.J.M. and W.D. van Hoorn (1983b) Prognose 1980, gevoeligheidsanalyse van het rekenmodel, *Maandstatistiek van de Bevolking* 31(12), 20-30.

Cruijsen, H.G.J.M. and N.W. Keilman (1989) National population forecasts in industrialized countries: Main findings of an international workshop, *Maandstatistiek van de Bevolking* 37(11), 19-23.

Dammers, E. (1990) Trefzekerheidsmaten voor de evaluatie van kwantitatieve voorspellingen, in: H.A. Becker and G. Dewulf (eds.) *Terugkijken op toekomstonderzoek*, ISOR/ University of Utrecht, 351-399.

D'Andrade, R. (1986) Three scientific world views and the covering law model. In: D.W. Fiske and R.A. Shweder (eds.), *Metatheory in social science: Pluralism and subjectives*, The University of Chicago Press, Chicago and London, 19-41.

Davis, K. (1979) The continuing demographic revolution in industrial societies, in: S. Lipset (red.), *The third century: America as a post-industrial society*, University of Chicago Press, Chicago, 37-64.

Dasgupta, A.K. and D.W. Pierce (1978) *Cost benefit analysis: Theory and practice*, The MacMillan Press Ltd., London etc.

De Beer, J. (1985) A time series model for cohort data, *Journal of the American Statistical Association* 80(391), 525-530.

De Beer J. (1987) Accuracy of short-term fertility forecasts, paper European Population Conference Jyväskylä, Finland, June 11-16.

De Beer, J. (1988) Predictability of demographic variables in the short run, *European Journal of Population* 4(4), 283-296.

De Feijter, H. (1989) Voorlopers en navolgers bij demografische veranderingen, unpublished paper Planologisch en Demografisch Instituut, Universiteit van Amsterdam.

De Jong, A. (1987) Het eerste en tweede geboorte-interval nader belicht, *Maandstatistiek van de Bevolking* 35(5), 14-23.

Dennis R.L. (1987) Forecasting errors: The importance of the decision-making context, *Climatic Change* 4, 81-96.

Dorn, H. (1950) Pitfalls in population forecasts and projections, *Journal of the American Statistical Association* 45, 311-334.

Ekamper, P. (1990) Gevoeligheidsfuncties voor meerdimensionale bevolkingsprognose-modellen met een twee-geslachten algoritme, NIDI-Rapport no. 14, Netherlands Interdisciplinary Demographic Institute, The Hague, The Netherlands.

Feichtinger, G. (1971) Stochastische Modelle demographische Prozesse, Lecture Notes in Operations Research and Mathematical Systems no. 44, Springer Verlag, Berlin.

Feichtinger, G. and G. Sorger (1990) Capital accumulation, aspiration adjustment, and population growth: Limit cycles in an Easterlin-type model, *Mathematical Population Studies* 2(2), 93-103.

Field, J.L. (1988) Monitoring of the United Kingdom projections, Handout International Workshop on National Population Projections, Voorburg, The Netherlands, October 1988.

Fienberg, S.E. and W.M. Mason (1978) Identification and estimation of age-period-cohort models in the analysis of discrete archival data, in: K.F. Schuessler (ed.) *Sociological Methodology* 8, Jossey-Bass, San Francisco, 1-67.

Fildes, R. (1982) Forecasting: the issues, in: S. Makridakis and S.C. Wheelwright (eds.), *The handbook of forecasting: a managers guide*, John Wiley and Sons, New York etc., 83-104.

Finsterbusch, K. and A.B. Motz (1980) *Social Research for Policy Decisions*, Wadsworth Publishing Company, Belmont.

Frauenthal, J. and K. Swick (1983) Limit cycle oscillations of the human population, *Demography* 20, 285-298.

Frinking, G. and J. Godefroy (1967) De groei van de Nederlandse bevolking tot het jaar 2000, in: Ethiek van een Nederlandse bevolkingspolitiek, *Annalen van het Thijmgenootschap*, 15-20.

Gill, R. and N. Keilman (1989) On the estimation of multidimensional demographic models with population registration data, *Mathematical Population Studies* 2(2), 119-143.

Gleick, J. (1987) *Chaos: Making a new science*, Penguin Books, New York etc.

Glenn, N.D. (1976) Cohort analysts' futile quest: statistical attempts to separate age, period and cohort effects, *American Sociological Review* 41, 900-904.

Gordijn, H. and H. Heida (1979) Een multi-regionaal demografisch model en de ontwikkeling van een monitoringssysteem, Planologisch Studiecentrum TNO, Delft, The Netherlands.

Grauman, J. (1967) Success and failure of the 1950's: A general appraisal, Proceedings of the World Population Conference 1965, Vol. 3, United Nations, New York.

Hagenaars, J.A.P. (1985) Loglineaire analyse van herhaalde surveys: Panel-, trend- en cohortonderzoek, Ph.D. Thesis University of Tilburg, the Netherlands.

Hämäläinen, H. (1987) The accuracy of the regional population projections in

Finland in the 1980s, paper European Population Conference Jyväskylä, Finland, June 1987.
Henry, L. (1985) Voeux pour le XXIe siècle, International Population Conference Florence 1985, 5-12 June, Vol. 1, Liège, IUSSP.
Henry, L. (1987) Perspectives et prévision, in: Les projections démographiques, Actes du VIIIe Colloque National de Démographie, Tome 1, Travaux et Documents Cahier no. 116, Presses Universitaires de France, Paris, 3-11.
Henry, L. and H. Gutierrez (1977) Qualité des prévisions démographiques à court terme: Etude de l'extrapolation de la population totale des départements et villes de France, 1821-1975, *Population* 32(3), 625-647.
Henshel, R.L. (1982) Sociology and social forecasting, *Annual Review of Sociology*, 57-79.
Hertz, D.B. and H. Thomas (1983) *Risk analysis and its application*, Wiley, Chichester.
Heyde, C.C. and J.E. Cohen (1985) Confidence intervals for demographic projections based on products of random matrices, *Theoretical Population Biology* 27, 120-153.
Hoem, J.M. (1973) Levels of errors in population forecasts, Artikler fra Statistisk Sentralbyrå nr. 61, Oslo.
International Union for the Scientific Study of Population - IUSSP (not dated) Multilingual Demographic Dictionary: English Section, Ordina Editions, Liège.
Isserman, A. (1977) The accuracy of population projections for subcounty areas, *Journal of the American Institute of Planners* 43, 247-259.
Judge, G.C., W.E. Griffiths, R. Carter Hill, H. Lütkepohl and T.-S. Lee (1985) *The Theory and Practice of Econometrics*, Wiley, New York etc. (second edition).
Keilman, N.W. (1982) Bevolkingsprognoses: nu en in de jaren zestig, *Maandstatistiek van de Bevolking* 30(6), 12-23.
Keilman, N.W. (1985a) Internal and external consistency in multidimensional population projection models, *Environment and Planning A* 17, 1473-1498.
Keilman, N.W. (1985b) Nuptiality models and the two-sex problem in national population forecasts, *European Journal of Population* 1,207-235.
Keilman, N.W. (1986) The unpredictability of population trends, in: H.A. Becker and A.L. Porter (eds.) *Methods and experiences in impact assessment*, D.Reidel Publ. Co., Dordrecht etc.
Keilman, N. (1990) Methods of population projections in industrialized countries, forthcoming in W. Lutz (ed.) *Future demographic trends in Europe and North America.*
Keilman, N. and T. Kucera (forthcoming) The impact of forecasting methodology on the accuracy of national population forecasts: evidence from The Netherlands and Czechoslovakia (accepted for publication, *Journal of Forecasting*).
Keilman, N. and H. Cruijsen, eds. (forthcoming) *National Population Forecasting in Industrialized Countries.*
Keyfitz, N. (1972) On future population, *Journal of the American Statistical Association* 67, 347-363.
Keyfitz, N. (1977) *Applied mathematical demography*, J.Wiley and Sons, New York etc.
Keyfitz, N. (1981) The limits of population forecasting, *Population and Development Review* 7(4), 579-593.

Keyfitz, N. (1982) Can knowledge improve forecasts? *Population and Development Review* 8(4), 729-751.

Keyfitz, N. (1984a) The social and political context of population forecasting, IIASA Working Paper no. WP-84-3, International Institute for Applied Systems Analysis, Laxenburg, Austria.

Keyfitz, N. (1984b) Population projections as an aid to the formulation and implementation of population policies, in: *Population projections: Methodology of the United Nations*, Population Studies no. 83, United Nations, New York, 17-20.

Keyfitz, N. (1985a) Demography in the 21st century: the uses of forecasting, International Population Conference, Florence, 1985, Vol. 1, Liège, IUSSP, 59-81.

Keyfitz, N. (1985b) A probability representation of future population, *Zeitschrift für Bevölkerungswissenschaft* 11(2), 179-191.

Kregting, J.W.M. (1979) De bevolkingsvooruitberekening getoetst, *Maandstatistiek van Bevolking en Volksgezondheid*, 27(7) 245-251.

Kregting, J.W.M. (1980) De bevolkingsvooruitberekening getoetst, 1975-1979, *Maandstatistiek van Bevolking en Volksgezondheid* 28(9), 41-51.

Kuijsten, A.C. (1984) Measuring the accuracy of population projections: an evaluation of Keyfitz's concept "quality of prediction", NIDI Working Paper no. 49, Netherlands Interuniversity Demographic Institute, Voorburg, The Netherlands.

Kuijsten, A.C. (1989) Demografische toekomstbeelden van Nederland, Verkenningen no. 49 Planologisch en Demografisch Instituut, Universiteit van Amsterdam.

Laaksonen, S. (1980) On sensitivity in population forecasts, *Yearbook of population research in Finland* 18, 96-101.

Land, K. (1986) Methods of national population forecasts: a review, *Journal of the American Statistical Association* 81(396), 53-77.

Land, K.C. and S.H. Schneider (1987) Forecasting in the social and natural sciences: an overview and analysis of isomorphisms, *Climatic Change* 11(7), 7-31.

Le Bras, H. (1974) Populations stables aléatoires, *Population* 29, 435-464.

Lee, D.B. (1973) Requiem for large-scale models, *Journal of the American Institute of Planners* 39, 163-178.

Lee, R.D. (1974) Forecasting births in post-transition populations: Stochastic renewal with serially correlated fertility, *Journal of the American Statistical Association* 69, 607-617.

Leeson, G.W. (1981) The elderly in Denmark: consequences of a mortality decline, *European Demographic Information Bulletin* 12(3), 89-100.

Long, J.F. (1984) U.S. national population projection methods: a view from four forecasting traditions, *Insurance: Mathematics and Economics* 3, 231-239.

Long, J.F. (1985) Integrating multiple forecasting traditions in the Census Bureau's national population projections methodology, Proceedings First Annual Research Conference, U.S. Bureau of the Census, Washington D.C., 237-250.

Long, J.F. (1987) The accuracy of population projection methods of the U.S. Census Bureau, Paper annual meeting of the PAA, Chicago, April/May 1987.

Long, J.F. (1988) Accuracy, monitoring and evaluation of national population projections, paper International Workshop on National Population Projections, Voorburg, The Netherlands, October 1988.

Long, J.F. and D.B. McMillen (1987) A survey of Census Bureau population projection methods, *Climatic Change* 11, 141-177.

Macura, M. (1968) The long-range outlook - summary of current estimates, in: R.N. Farmer, J.D. Long and G.J. Stolnitz (eds.) *World population: The view ahead*, Bureau of Business Research, Graduate School of Business, Indiana University, 15-48.

Maddala, G.S. (1983) *Econometrics*, Mc-GrawHill International Book Company, Auckland etc. (5th printing 1983).

Mahmoud, E. (1984) Accuracy in forecasting: a survey, *Journal of Forecasting* 3(2), 139-159.

Makridakis, S. (1986) The art and science of forecasting: an assessment and future directions, *International Journal of Forecasting* 2, 15-39.

Makridakis, S. and M. Hibon (1979) Accuracy of forecasting: an empirical investigation (with discussion), *Journal of the Royal Statistical Society* A 142(2), 97-145.

Marini, M.M. and B. Singer (1988) Causality in the social sciences, in: C. C. Clogg (ed.) *Sociological Methodology* 18, 347-409.

Mason, W.M. and S.E. Fienberg (1985) *Cohort analysis in social research: Beyond the identification problem*, Springer Verlag, New York.

McDonald, J. (1979) A time series approach to forecasting Australian total live births, *Demography* 16, 575-601.

McDonald, J. (1981) Modeling demographic relationships, an analysis of forecast functions for Australian births, *Journal of the American Statistical Association* 76(376), 782-792.

Mishan, E.J. (1979) *Cost-benefit analysis: An informal introduction*, Allen and Unwin, London.

Moors, H.G. (1974) *Child Spacing and Family Size in the Netherlands*, Stenfert Kroese, Leiden, the Netherlands.

Müller, A. (1982) Einschätzung von Fehlerrisken demographischer Prognosen-Ergebnisse von Prognose/Realisations-Analysen, *Statistische Rundschau für das Land Nordrhein-Westfalen* 34(9), 499-508.

Müller, A. (1987) Zur Prognosegüte von Bevölkerungsprognosen: Prognose/Realisationsuntersuchungen mit dem regionalen Prognosemodell in Nordrhein-Westfalen, *Zeitschrift für Bevölkerungswissenschaft* 13(3), 345-378.

Murdock, S.H., F.L. Leistritz, R.R. Hamm, S.S. Hwang and B. Parpia (1987) An assessment of the accuracy of a regional economic-demographic projection model, *Demography* 21, 383-404.

Murphy, M.J. (1984) Population forecasting: alternative approaches, Office of Population Censuses and Surveys Occasional Paper.

Myers, R.J. (1954) Comparison of population projections with actual data, Proceedings World Population Conference 1954, Vol. III, United Nations, New York, 101-111.

Netherlands Central Bureau of Statistics - NCBS (1951) Berekeningen omtrent de toekomstige loop der Nederlandse bevolking, Uitg. Mij. W. de Haan N.V., Utrecht.

NCBS (1954) Herziene berekeningen omtrent de toekomstige loop der Nederlandse bevolking 1951-1981, Uitg. Mij. W. de Haan N.V., Utrecht.

NCBS (1959) Vergelijking van de werkelijke met de vooruitberekende loop van de Nederlandse bevolking, *Statistische en Econometrische Onderzoekingen* 1e kwartaal 1954.

NCBS (1961) Berekeningen omtrent de toekomstige loop van de Nederlandse bevolking, *Maandschrift van het Centraal Bureau voor de Statistiek*, Jaargang 56, 657-661.
NCBS (1965a) Berekeningen over de groei van de Nederlandse bevolking in de periode 1965-1980, *Maandstatistiek van Bevolking en Volksgezondheid* 13(2), 41-55.
NCBS (1965b) Berekeningen over de groei van de Nederlandse bevolking in de periode 1965-2000, *Maandstatistiek van Bevolking en Volksgezondheid* 13(4), 109-111.
NCBS (1967a) Aanvullende berekeningen over de Nederlandse bevolkingsgroei tot 1980 naar aanleiding van de na 1964 opgetreden geboortedaling, *Maandstatistiek van Bevolking en Volksgezondheid* 15(4), 107-111.
NCBS (1967b) Berekeningen over de Nederlandse bevolkingsgroei tot 2000 naar aanleiding van de na 1964 opgetreden versnelde geboortedaling, *Maandstatistiek van Bevolking en Volksgezondheid* 15(9), 293-299.
NCBS (1971) Berekeningen omtrent de toekomstige bevolkingsgroei in Nederland in de periode 1970-2000, Staatsuitgeverij, The Hague.
NCBS (1973) De toekomstige Nederlandse bevolkingsontwikkeling na 1972, Staatsuitgeverij, The Hague.
NCBS (1974a) Huwelijksvruchtbaarheid: Een cohortanalyse 1937-1971, Staatsuitgeverij, The Hague.
NCBS (1974b) Onderzoek Gezinsvorming mei 1974, *Maandstatistiek van Bevolking en Volksgezondheid* 22(12), 366-376.
NCBS (1975) Jaarverslag van het Centraal Bureau voor de Statistiek over 1975.
NCBS (1976) De toekomstige demografische ontwikkeling in Nederland na 1975, Staatsuitgeverij, 's-Gravenhage.
NCBS (1977) Supplement Jaaroverzicht bevolking en volksgezondheid 1975, *Maandstatistiek van Bevolking en Volksgezondheid* jaargang 24.
NCBS (1979) De toekomstige ontwikkeling van de burgerlijke staat van mannen, 's-Gravenhage, Staatsuitgeverij.
NCBS (1982) Prognose van de bevolking van Nederland na 1980. Deel I: Uitkomsten en enkele achtergronden, Staatsuitgeverij, The Hague.
NCBS (1984) Prognose van de bevolking van Nederland na 1980. Deel II: Modelbouw en hypothesevorming, Staatsuitgeverij/CBS-publikaties, The Hague.
NCBS (1985) Bevolkingsprognose voor Nederland 1984-2030, Staatsuitgeverij/ CBS-publikaties, The Hague.
NCBS (1988) Onderzoek Gezinsvorming, februari 1988, *Maandstatistiek van de Bevolking* 36(11), 46-47.
NCBS (1989) Bevolkingsprognose voor Nederland 1988-2050, SDU-uitgeverij/CBS-publikaties, The Hague.
NCBS (1990) Uitkomsten bevolkingsprognose 1989, *Maandstatistiek van de Bevolking* 38(1), 23-38.
Newbold, P. and C.W.J. Granger (1974) Experience with forecasting univariate time series and the combination of forecasts, *Journal of the Royal Statistical Society A* 137, 131-165.
OECD (1961) Demographic trends 1956-1976 in Western Europe and in the United States, OECD, Paris.
OECD (1966a) Demographic trends: supplement country reports, OECD, Paris.
OECD (1966b) Demographic trends 1965-1980 in Western Europe and North

America, OECD, Paris.
Olshansky, S.J. (1988) On forecasting mortality, *The Milbank Quarterly* 66(3), 1-49.
Openshaw, S. and G.A. van der Knaap (1983) Small area population forecasting: Some experience with British models, *Tijdschrift voor Economische en Sociale Geografie* 74(4), 291-304.
Op 't Veld, D., E. Bijlsma, G. Eshuis and J. Bosten (1984) Goodness-of fit? Een aantal goodness-of-fit maten op hun gedrag bekeken, Planologisch Studiecentrum TNO, Delft.
Page, W. (1982) Long-term forecasts and why you will probably get it wrong, in S. Makridakis and S.C. Wheelwright (eds.) *The Handbook of Forecasting: A Manager's Guide*, J. Wiley and Sons, New York etc., 449-456.
Pflaumer, P. (1984) Die Berücksichtigung der Unsicherheit bei der zukünftigen Entwicklung der Bevölkerung und der Renten-beitragsatze in der Bundesrepublik Deutschland, *Zeitschrift für Bevölkerungswissenschaft* 10(4), 501-530.
Pflaumer, P. (1988) Confidence intervals for population projections based on Monte Carlo methods, *International Journal of Forecasting* 4, 135-142.
Pflaumer, P.(1988b) *Methoden der Bevölkerungsvorausschätzung unter besonderer Berücksichtigung der Unsicherheit*, Volkswirtschaftliche Schriften Heft 377, Duncker und Humblot, Berlin.
Pittenger, D.B. (1977) Population forecasting standards: some considerations concerning their necessity and content, *Demography* 14(3), 363-368.
Pollard, J.H. (1966) On the use of the direct matrix product in analyzing certain stochastic population models, *Biometrika* 53, 397-415.
Pressat, R. (1975) Mortality projections and actual trends: A comparative study, *Tijdschrift voor Sociale Geneeskunde* 53, 51-62 (also appeared in *World Health Statistics Report* 27 (1974) 516).
Pressat, R. (1983) *L'Analyse Démographique: Concepts - Méthodes - Résultats*, Presse Universitaire de France, Paris (Quatrième Edition).
Preston, S.H. (1974) An evaluation of postwar mortality projections in Australia, Canada, Japan, New Zealand and the United States, *WHO Statistical Report* 27, 719-745.
Rademaker, P. (1979) Toekomstverkenning: Een plaatsbepaling en een praktijkvoorbeeld, in: J.W. van der Ben, M.T.G. Meulenberg, A.P.N. Nauta, W.F. van Raaij, G.J. Randoe, J.J.M. van Tulder, A. van der Zwan (eds.) *Marktonderzoek en Consumentengedrag*, Jaarboek 1979 van de Nederlandse Vereniging van Marktonderzoekers, Amsterdam, 95-113.
Reynolds, H.T. (1977) *The Analysis of Cross-Classifications*, The Free Press, New York and London.
Rideng, A. (1988) The evaluation of regional population problems, in: O. Bertelsen and B. Liebach (eds.) *The eighth Nordic Demographic Symposium, June 1986, Gilleleje, Denmark*, Scandinavian Population Studies 8, The Nordic Demographic Society, Copenhagen, 265-279.
Rodgers, W. (1982) Estimable functions of age, period and cohort effects, *American Sociological Review* 47, 774-787.
Rogers, A. (1975) *Introduction to multiregional mathematical demography*, Wiley, New York etc.
Sabatello, E.F. (1988) Population ptojections of the Adriatic populations toward the 21th century, in: IDRA I, Atti del primo Incontro Demografico delle

Regioni Adriatiche (Pescara, novembre 1987), Istituto di Statistica, Università "G. D'Annunzio", Pescara, 283-291.
Saboia, J. (1974) Modeling and forecasting populations by time-series: The Swedish case, *Demography* 11, 483-492.
Saboia, J. (1977) Autoregressive integrated moving average (ARIMA) models for birth forecasting, *Journal of the American Statistical Association* 72, 264-270.
Scheéle, S. (1981) Osäkerhet i befolkningsprognoser, Stockholms Kommune Utrednings - och Statistikkontor, Stockholm.
Schreurs, S. (1989) Een beschrijvende analyse van de trefzekerheid van nationale bevolkingsprognoses, NIDI-rapport no. 2, Netherlands Interdisciplinary Demographic Institute, The Hague.
Schweder, T. (1971) The precision of population projections studied by multiple prediction methods, *Demography* 8(4), 441-450.
Shweder, R.A. and D.W. Fiske (1986) Introduction: Uneasy Social Science, In: D.W. Fiske and R.A. Shweder (eds.). Metatheory in social science: Pluralism and subjectives, The University of Chicago Press, Chicago and London, 1-18.
Siegel, J.S. (1972) Development and accuracy of projections of population and households in the United States, *Demography* 9, 51-68.
Smith, S.K. (1984) *Confidence intervals for county population forecasts*, paper annual meeting of the PAA, Minneapolis, May 1984.
Smith, S.K. (1987) Tests of forecast accuracy and bias for country population projections, *Journal of the American Statistical Association* 82(400), 991-1003.
Smith, S.K. and T. Sincich (1988) Stability over time in the distribution of population forecast errors, *Demography* 25(3), 461-474.
Smulders, R. and R. Verhoef (1973) Het rekenschema van de Nederlandse bevolkingsprognose in 1973, *Bevolking en Gezin* 1973, 3, 389-426.
Somermeyer, W.H. (1961) Een analyse van de binnenlandse migratie in Nederland tot 1947 en van 1948-57, *Statistische en Econometrische Onderzoekingen*, 115-174.
State Commission for Population Issues (1977) *Bevolking en welzijn in Nederland*, Staatsuitgeverij, The Hague.
Statistiska Centralbyrån - SCB (1986) Den framtida befolkningen: Prognos för åren 1986-2025, Demografiska Rapporter, Sveriges Officiella Statistik, Statistiska Centralbyrån, Stockholm.
SCB (1989) Sveriges framtida befolkning: Prognos för åren 1989-2025, Demografiska Rapporter 1989:1, Sveriges Officiella Statistik, Statistiska Centralbyrån, Stockholm.
Stoto, M.A. (1983) The accuracy of population projections, *Journal of the American Statistical Association* 78, 13-20.
Stoto, M.A. (1988) Dealing with uncertainty: statistics for an aging population, *The American Statistician* 42(2), 103-110.
Stoto, M.A. and A.P. Schrier (1981) The accuracy of state population projections, John Fitzgerald Kennedy School of Government Discussion Paper Series no. 117 D, Harvard University.
Suppes, P. (1984) *Probabilistic Metaphysics*, Basil Blackwell, Oxford.
Sykes, Z.M. (1969) Some stochastic versions of the matrix model for population dynamics, *Journal of the American Statistical Association* 64, 111-130.
Theil, H. (1966) *Applied economic forecasting*, Rand Mc-Nally, Chicago.
Timmermans, H. and A. Borgers (1984) On the assessment of model performance in spatial choice analysis, paper Department of Architecture, Building and

Planning, University of Technology, Eindhoven, the Netherlands.
United States Bureau of the Census - USBC (1984) Projections of the population of the United States by age, sex and race: 1983-2080, Current Population Reports, Series P-25 no. 952.
USBC (1989) Projections of the population of the United States by age, sex and race: 1988-2080, Current Population Reports, Series P-25 no. 1018 (by G. Spencer).
Vallin, J. (1989) L'avenir de l'espérance de vie vu à travers les projections de l'INSEE, *Population* 44 (4-5), 930-936.
Van Dantzig, D. (1952) Voorspelling en profetie, *Statistica (Neerlandica)* 6(4), 195-203.
Van de Giessen, G.J. (1984) De betrouwbaarheid van vruchtbaarheidsverwachtingen in Nederland, *Maandstatistiek van de Bevolking* 32(3), 42-55.
Van de Giessen, G.J. (1985) The use of NFS'82 data in the 1984-based population forecasts for the Netherlands, paper, Netherlands Central Bureau of Statistics, Voorburg (revised 1987).
Van de Giessen, G.J. (1987a) De betrouwbaarheid van de verwachting binnen drie jaar (g)een kind te krijgen, *Maandstatistiek van de Bevolking* 35(7), 13-20.
Van de Giessen, G.J. (1987b) De stabiliteit van lange termijn geboorteverwachtingen 1982-1985, *Maandstatistiek van de Bevolking* 35(11), 20-29.
Van de Giessen, G.J. and W.D. van Hoorn (1985) Bevolkingsprognose 1984: hypothesen betreffende de geboorte, *Maandstatistiek van de Bevolking* 33(8), 28-46.
Van de Mortel, P. (1988) Inventarisatie en evaluatie vam nationale bevolkingsprognoses, Intern Rapport no. 55, Netherlands Interuniversity Demographic Institute, The Hague.
Van Imhoff, E. (1988) A general characterization of consistency algorithms in multistate projection models, Paper prepared for presentation at the International Workshop on Multistate Demography, Zeist, The Netherlands, October 31 - November 4, 1988.
Van Imhoff, E. (forthcoming) The exponential multidimensional demographic projection model, *Mathematical Population Studies*.
Van Nimwegen, N. (1981) Developments in Dutch Demography: an introduction, in: H.G. Moors (ed.) *National Population Biography of the Netherlands 1945-1979*, Netherlands Interuniversity Demographic Institute, Voorburg, the Netherlands.
Van Poppel, F. (1984) Demografie 1964-1984, *I.D.G.-Bulletin* 1983/84, 7-15.
Van Seventer, H. and A.P. Vossen (1979) Enkele theoretische en methodologische grondslagen van de bevolkingsprognostiek: Een eerste verkenning, Katholieke Hogeschool Tilburg, Subfaculteit Sociaal-Culturele Wetenschappen.
Voss, P.R, C.D. Palit, B. Kale and H.C. Krebs (1981) Population projections for states: An analysis of error structure, in: American Statistical Association, 1981 Proceedings of the Social Statistics Section, 436-441.
Vossen, A.P.G.J. (1988) De bevolkingsvooruitberekening: Op zoek naar nieuwe wegen, Ph.D. Thesis Catholic University of Brabant.
Wachter, K.W. and R.D.Lee (1989) U.S. Births and limit cycle models, *Demography* 26(1), 99-115.
Wetenschappelijke Raad voor het Regeringsbeleid - WRR (1974) Bevolkingsprognose, 's-Gravenhage, Staatsuitgeverij.
Whelpton, P.F. (1928) Population of the United States, 1925 to 1975, *The American Journal of Sociology* 34(2), 253-272.

White, H.R. (1954) Empirical study of the accuracy of selected methods of projecting state populations, *Journal of the American Statistical Association* 29, 480-498.

Willekens, F.J. (1977) Sensitivity analysis in multiregional demographic models, *Environment and Planning A* 9, 653-674.

Willekens, F.J. (1984) Spatial policy and demographic research opportunities, in: H. ter Heide and F.J. Willekens (eds.) *Demographic research and spatial policy: The Dutch experience*, Academic Press, London etc., 355-401.

Willekens, F. and N. Baydar (1984) Age-period-cohort models for forecasting fertility, NIDI-Working Paper no. 45, Netherlands Interuniversity Demographic Institute, Voorburg, the Netherlands.

Willekens, F.J. and P. Drewe (1984) A multiregional model for regional demographic projection, in: H. ter Heide and F.J. Willekens (eds.), *Demographic research and spatial policy: The Dutch experience*, Academic Press, London etc., 309-334.

Williams, W.H. and M.L. Goodman (1971) A simple method for the construction of empirical confidence limits for economic forecasts, *Journal of the American Statistical Association* 66(366), 752-754.

Author Index

Subject Index

Italics indicate pages where definitions are given